Youth in Conflict and Peace Building in North East India

The Author

Dr. Haans J. Freddy is Assistant Professor in the Department of Political Science, Madras Christian College, Chennai. He was earlier a Postdoctoral fellow with the Indian Council of Social Science Research where he researched on youth rights in north east India. His research interests are on international relations, strategic studies, human rights, political thought and security studies. He was also a fellow of the Japan International Cooperation Agency. He is a member of the editorial board in the Future of Food: Journal of Food Agriculture and society which is published by the University of Kassel, Germany.

Youth in Conflict and Peace Building in North East India

— *Author* —

Haans J. Freddy

Department of Political Science, Madras Christian College
Chennai – 600 059, Tamil Nadu

2021

Regency Publications

A Division of

Astral International Pvt. Ltd.

New Delhi – 110 002

ISBN : 9789390371884 (Int. Edn.)

Published by : **Regency Publications**
A Division of
Astral International Pvt. Ltd.
– ISO 9001: 2015 Certified Company –
4736/23, Ansari Road, Darya Ganj
New Delhi-110 002
Ph. 011-43549197, 23278134
E-mail: info@astralint.com
Website: www.astralint.com

For

K'Liu

Acknowledgements

I would like to express my sincere gratitude and thanks to the LORD JESUS for his love and support during a very difficult phase of my life that I went through while writing this book.

My sincere thanks to Dr. C. Joshua Thomas, Deputy Director, Indian Council of Social Science Research-North Eastern Regional Council, Shillong, for all his kind words of encouragement. Dr. Thomas has always been a source of inspiration and it is through his guidance and inspiration that I had written this volume.

A special mention has to be made of Dr. Lawrence Prabhakar Williams, Associate Professor, Department of Political Science, Madras Christian College, who encouraged and supported me immensely while writing this book. I am grateful for the kind response of my friends as well as those strangers I met while writing this book. I would like to especially thank Dr. K. Debbarma, Dr. R. Borgohain, Prof. B.J. Deb, Dr. Biju Kumar, Dr. Hariharan, Dr. K. Palani and Dr. S.D. Christopher Chandran, Dr. R. Sridhar and my friend and colleague Dr. Ashik Jacob Bonofer.

A sincere word of thanks to Dr. Allen J. Freddy for his kind words of encouragement, love and support. My Mother, who was a great source of support and encouragement, needs a special note of thanks.

I extend my sincere thanks to my father, late Dr. K John Freddy, who was my source of inspiration to begin a career in academics.

Words do not suffice to express my sincere gratitude to my wife K'Liu, who supported and encouraged me through the long hours of work away from home.

And to my sons Raphael and Zacchaeus, who have always been a source of happiness in difficult times – a big thank you.

Haans J. Freddy

Preface

Until the end of the cold war, the conventional wisdom in the world was that ethnicity and nationalism were outdated concepts and largely resolved problems. On both sides of the cold war, the trend seemed to indicate that the world was moving toward internationalism rather that nationalism. As a result of the threat of nuclear warfare, great emphasis on democracy and human rights, economic interdependence and gradual acceptance of universal ideologies, it became fashionable to speak of the demise of ethnic and nationalist movements. Despite expectations to the contrary, however, a fresh cycle of ethno-political movements have re-emerged recently in eastern Europe, central Asia, Africa and South Asia and many other parts of the world. Infact with the end of the cold war, which clearly increased international cooperation while decreasing the possibilities of inter-state wars, the main threat to peace does not come from major inter-state confrontations anymore, but from another source – intra-state conflicts or conflicts that occur within the borders of states. These conflicts have replaced the cold war's ideological clashes as the principal sources of current conflict. The fact that internal conflicts occur within the borders of states made major international actors reluctant to intervene as well, either for legal concerns or concern to avoid probable loses. Thus unless conflicts really escalate, the international community has preferred not to get involved in intra-state conflicts. But such conflicts are as serious, costly and intense as any in the past. These conflicts somehow need to be managed and reduced, or else international peace and security will not be in a stable situation. Even if intra-state conflicts appear to be local, they can quickly gain an international dimension due to global interdependence and to various international support. Infact, when external parties provide political, economic, military assistance, asylum or the bases for actors involved in local struggles, these conflicts inevitably assume an international dimension. Undoubtedly, the effective management of intra-state conflicts by domestic and international authorities presupposes an understanding of their nature and causes. Others in contrast have noted that the post-World War II era has witnessed a great deal of warfare. Most brutally, tens of millions have died in combat in the modern era and even more in the acts of genocide, politicide and democide that often accompany war and are exacerbated by it. There are those who claim that both international and civil wars have made a comeback. Other scholars have focused more specifically on the emergence of possibly new types of war. For instance, Kal Holsti has argued that we are seeing the emergence of new wars of the third kind and others decry the proliferation of ethnic wars. For all these reasons and more, it is vitally important to know whether war is ubiquitous as some say, declining as fast as others assert, or fluctuating across time and place. Surprisingly, though hundreds of books and articles have tangentially addressed this question in recent years, too few have examined and answered the basic question of whether the data indicate that the total amount of war in the world is increasing or decreasing or remaining the same. Research endeavors have tended to focus on only one type of war and many of the contradictory conclusions are primarily directed by the type of war being studied (inter-state or civil). Obviously part of this dichotomy is linked to cleavages in the discipline, with international politics scholars primarily concerned with wars between states and scholars in the field of comparative politics have their attention focused on internal developments that include revolutions and civil wars. The lack of a dialogue between these two realms has limited the accumulation of knowledge and has circumscribed our understanding of war. More particularly, following the end of the cold war, a narrow disciplinary focus exclusively on inter-state wars is no longer justifiable

Wars between states have declined sharply in number since the end of World War II along with the founding of the United Nations Organization. This is primarily the result of a generalized attempt to avoid further conflicts. War amongst people which is referred as intra-state war makes a paradigm shift in the approach to understanding conflicts. Intra-state wars mainly oppose the state to feudal, often warlike, organizations seek to usurp state prerogatives. To simplify, these feudal organizations can be terrorist groups, guerillas, organized crime syndicates or a combination of these types of organizations. The majority of these are under the domination of more powerful groups whose goal is the conquest of local or national sovereignty

Youth studies in recent times has received immense attention among policy makers, scholars activists and researchers. The concept of youth which is seen as being flexible in nature is debated by scholars both as a biological or a social construct. The idea that youth are the future generation, is one of the factors that has led many to think and research on the problems and issues that young people as a group or as individuals face in society. While there has been a proliferation of literature on the concept of youth, this volume considers the fact that young people are critical to the concept of conflict and peacebuilding. While examining the participation of youth in armed conflict, it examines a variety of debates that analyse the reasons for such engagement. There are various arguments that posits youth as active participants in conflict and there are political and economic reasons for such participation. It takes into consideration the theory of greed and grievances for young people to engage actively in conflicts. The argument that the presence of youth affects war and not how war affects conflicts is plausible.

This book also takes into consideration how young people can be engaged positively in any peacebuilding process and more particularly in the north east of India. While young people are considered to be the future generation there is also the need to consider the fact that as individuals belonging to the future, their needs and demands need to be taken into consideration for establishing lasting peace.

In this book, the conflict in north east India which has survived for over five decades has drawn a variety of responses in terms of analysing the conflict in the region. Much of the research that has been conducted until now have very little in terms of examining young peoples' participation in conflict and peacebuilding in the region. While there is the presence of student's unions in almost all eight states in north east India, there is however, little that has been done to include them in peace processes while on the contrary their participation in activism, armed conflict and violence is relatively very high. In this book, I posit that the engagement of youth in peacebuilding in north east India could bring lasting peace in the region that could result in change and bring development for the people to live a life with dignity, peace and harmony.

Ashik Jacob Bonofer

Contents

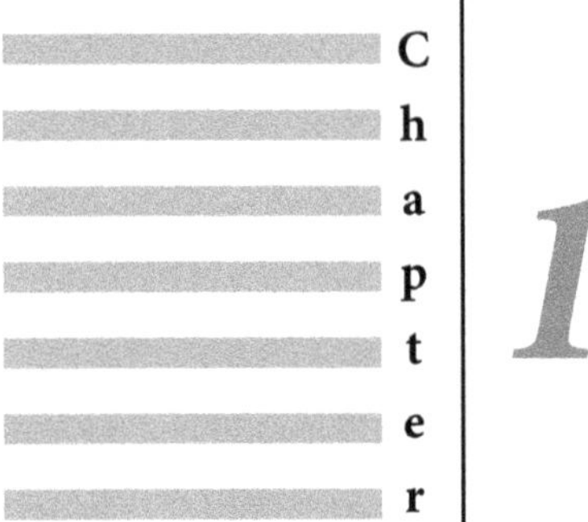

Introduction: The Concept of Youth

Over the last 40 years, young people have in the developed world been the subject of an enormous amount of research. Research on young people in general assumes that they are a significant category of people and constitute a separate group which we can term as non-adults. The problematic nature of being a young person and an even more problematic issue is the process of becoming an adult is a central and recurring theme in the study of youth. The study of youth and much of the available literature has inherited assumptions that are from the fields of developmental psychology that deals with the universal stages of development, identity formation, normative behaviour, and the relationship that is found between social and physical maturation. However, there has been very little work that has sought to clarify the theoretical basis of this categorization based on age. Allen (1996) made an argument that the concept of youth needs to be reassessed where she pointed out that it is not the relations in terms of ages that creates change and stability, but change in society that explains the relationship between different age groups. It was twenty years later that Jones (1998) pointed out that a need to develop a sociology of youth which would provide a conceptual framework and understanding the transitional stage that young people pass through as they become adults and also examining the different experiences of young people from different social groups was required. To emphasize on the qualities or otherwise of youth per se would be misleading, she argued, as young people are neither a homogeneous group nor a static one. According to her conclusion, youth is most usefully conceptualized as an age related process. This means that the focus on youth is not placed on the inherent characteristics of young people themselves, but on the construction of youth through social processes such as schooling, families or the labour market. Youth engage with these institution in specific ways in the context of their historical circumstances.

While there is growing awareness amongst contemporary researchers on the concept of youth, the focus on youth as a process throws into question the very use of the universal term 'youth'. For example, recently, Liebau and Chisholm (1993) have suggested that in Europe the concept or the idea of youth does not exist. Their argument is that each nation has its own cultures and economies that are directed by their own course and young people if different countries in Europe have to negotiate these differing circumstances accordingly vis-à-vis each other. They are shaped both by material objective aspects of cultures and societies in which they grow and by the ways in which they subjectively interpret their circumstances (Liebau and Chisholm, 1993). Wallace and Kovacheva (1995) while focussing on youth have pointed out that the experience of youth is being deconstructed because the significant transitions in life are less age related. In this context, it is interesting to note that the most significant issue which confronts the concept of youth is the apparent symmetry between biological and social processes. Age is a concept that is used to refer to the biological reality. However, the meaning and experience of age and the process of ageing is defined according to historical and cultural processes. Although every individual's life span can be measured objectively by the passing of time, it is cultural understandings about life stages that give the process of growing up and of ageing, its social meaning. It is through specific social and political processes that provide the frame within which cultural meanings are developed. Depending on the social, cultural and political circumstances, the terms youth and childhood have had and continue to have different meanings. Research on young people's lives in non-western countries exposes the ideal of the happy and safe childhood and period of youth as myths built around the social preoccupations and priorities of the capitalist countries of Europe and the United States (Boyden, 1990). In stark contrast to the western ideal of youth, she poses the trafficking and sexual exploitation of children in Thailand and the Philippines, the crimes committed against youth in Argentina during the time of the military regime and the detention of young people in South Africa under apartheid. More importantly, Boyden refers to the ethnographic studies which reveals how children and young people are expected to work for an income in some societies such as in India and Bangladesh which are not just for economic reasons but because of the belief that young people should engage with adult life as early as possible.

The idea of youth as socially constructed becomes more apparent when a global perspective is adopted. For a large proportion of the world's young people, the idea of youth as a universal stage of development was and continues to remain as an inappropriate concept. Although there is wide variation on the experiences of youth and in some instances where it may not even exist, it is yet important to understand some of the complexities of social change and the intersections that occur between institutions and personal biography. It can be argued that the concept of youth can be seen in the context of relational aspects which means or points to the social process by which age is socially constructed, institutionalized and controlled in historically and culturally specific ways. At this juncture it would be useful to refer to the earlier conceptual debates over the concept of gender since there are similarities. In the 1960s and 1970s, the idea of sex roles was a powerful concept that drew the attention towards analysing the inequalities that existed between men and women. The concept of sex roles provides the framework by which both men and women become limited by socially constructed categories or roles. Although this concept offered a useful descriptive model and significant basis for educational strategies to address gender inequality (Connell, 1987), it had serious setbacks. The static and categorical concept of sex roles, in which masculinity and femininity were seen as simply discrete, if constructed socially, these categories failed to provide a grasp in terms of the relationship between masculinities and femininities. This was finally replaced by a more sophisticated understanding of gender as a relational concept with power being placed at the centre. As a relational concept gender draws attention to the ways in which masculinity and femininity are constructed in relation to each other. They are not simply different categories and they cannot be understood independently of each other. For example, Davies, describes how boys in her study worked hard to maintain a dualism between boys and girls by denigrating and constantly drawing attention to feminine behaviour. This example provides a clear understanding in which being masculine involves maintaining a hierarchy where being male has greater value (Davies, 1993). In this context, there are several useful discussions of this conceptual issue where Franzway and Lowe (1978), Connell (1987) and Edwards (1983) have provided extensive analyses of the limitations of the categorical approach to gender relations.

Youth is understood as a relational concept as its exists and has its meaning largely in relation to the context of adulthood. The concept of youth as idealized and institutionalized directs our understanding about the eventual arrival at the status of adulthood. If youth is a state of becoming, adulthood is the arrival and at the same time youth is also not deficit of the adult state. One of the dimensions of the concept of youth is the position in which young people require guidance and expert attention so that their process of becoming adults is appropriately channelled. Understanding the concept of youth in the relational context, brings power relations to the forefront. This is an important aspect in terms of understanding the experiences that youth belonging to different groups have in the context of their growing up. The popular imaging of youth presenting a threat to law and order often represents young people as more powerful than they really are. Young people although they have rights as citizens, these can be relatively and easily denied and they also might have very little say in the institutions in which they have the most at stake – education (Aries, 1965).

The concept of youth in every day parlance is associated in a common sense manner with the state of being young more particularly with the phase of life that is between childhood and adulthood. The word youth in many occasions is used interchangeably with young person. Although it appears to mean the same thing, it however with the plural 'youths' the meaning broadens. The term 'youths' is a word which carries a great deal of baggage. This baggage includes ideas such as unruly young people – often male, operating in groups and the very least being a nuisance on the streets (Davies, 2004). Thus, the concept of youth although used with neutral descriptions is often not as we assume it to be. The concept when not used critically and carefully brings with it particularly negative assumptions in terms of the behaviour and character of young people both as individuals and in groups. Being young refers to the natural biological phase in the life cycle that is associated with the growth from childhood to adulthood. While it is connected with biological underpinnings, it is connected with the society as well which means that it has both social and biological meanings. People grow within a particular social context and young people occupy particular places within any given society. The experiences of young people and meanings that are attached to the term are derived directly from the social and economic positions that they occupy as young people as much as from their biological development. Taken in this sense the meaning of youth, the baggage which it carries differs and changes in time and place. Being young in the past was experienced and understood differently vis-à-vis being young today and when seen from this perspective the same may be the case in terms of being young in one part of the world that carries along with it different implications from being young in another (Eisenstadt, 1956).

Thus, youth in terms of a social concept includes both historical and spatial dimensions. Individuals and groups occupy different social positions and play different roles and society is structured in that manner within any given society. Social structure in general has an impact on the distribution of wealth and power and it is this distribution that brings imbalances among different groups. In the context of youth, being a young prince brings with it an entirely different status and identity, different social behaviour, expectations and opportunities vis-à-vis a young man who earns a minimum wage through some form of employment. Therefore, even though it is possible to identify certain biological characteristics that are common, it is also true that there is no one universal set of meanings that young people can fit into.

Nevertheless, the meanings that are attached to the concept of youth and the common usage of the term points to the dominant attitudes that are directed towards young people. These in turn has an impact on how young people are in general treated and perceived. Youth in general are affected by differences in access to wealth and resources that is partly related to legal age barriers which define access to social opportunities such as voting, employment and welfare and housing benefits and also related to the notion that youth is a period of learning, apprenticeship and a period of training to become an adult (Irwin, 1995).

The concept of youth is one that suggests similarities amongst people who belong to a particular age grouping which is used for the purpose of creating social rules and institutions which reinforces such similarities. This affects how young people understand and interpret their being young. Therefore, youth is a real social as well as a biological experience. However, because at the same time, realities of life for different individuals and different groups of young people are varied in terms of wealth or power defined by different categories such as class gender or citizenship, there is no possibility of a universal experience of youth. Such an understanding of the complex relationship that exists between the idea or the concept of youth and the different realities of young people's lives can help in understanding of the world which different young people inhabit (Wallace and Cross, 1990).

A definition of young people is required in order that we know exactly who it is that we are researching, working with or developing a policy for. As it is noted elsewhere in this book definitions of youth often reflect the biases of those who are doing the defining. Many of these assumptions starting from sociologists, youth workers and policy makers will reflect their own differing notions of what constitutes young people and many of these assumptions differ with the ways young people see themselves (Mannheim, 1952). Similarly, writers also document how conceptions of the period of youth are specific in terms of historical and cultural facts. In the western conception, the idea of youth is a relatively recent phenomenon starting from the eighteenth century (Gillis, 1974) although the discussion on the notion of youth can be traced back to the classical Greek society. Associated with the rise of western modernity, the idea of an intermediate stage of life between childhood and adulthood became important. Prior to the 1800s, childhood was seen to merge into an early form of adult independence from the ages of eleven to twelve as children took on some form of employment and assumed greater responsibilities and duties around their homes (Gillis, 1974).

Over the past two centuries in western societies, the historical and gradual emergence of the extension of the phase of youth as the socio-cultural definitions of dependent childhood and independent adulthood have become more clearly demarcated producing the notion of the youth phase as an interstitial phenomenon – existing between dependency of childhood and autonomy of adulthood. It was during the nineteenth and the twentieth century that the development of adult citizenship rights (Marshall and Bottomore, 1997) that helped define the many facets of the transition from childhood to adulthood and with it the contours of the career routes and status passages that young people have to travel to achieve adult independence. (Coles, 1995; Jones and Wallace, 1992).

During the late nineteenth century, in western societies, where the heralding of the end of child labour and the separation of employment from the domestic sphere were documented by commentators showed the common characteristics and experiences of young people. G.S. Hall (1904) and Erickson (1968) for example, showed the developmental characteristics of the phase of youth and the inevitable 'storm and stress' that accompanied this period of identity formation and movement through status passages to adulthood. Key to these assumptions of adolescence was the notion that the period of youth represented a time of flux where young individuals had time and space to experiment with their ideas as well as of others and identities as well as the actual routes that they may opt through life. Yet it must be noted that young people who lived in the twentieth century were conditioned heavily by class, race, and gender that defined much of their early lives as to what they might become in adult life. Influential writers such as Mannheim (1952) and Parsons (1942) in the twentieth century noted that young people recognized their common way of life (ideas, culture and life chances) and what they shared with other youth vis-à-vis adult society which set up the possibility of generational conflicts and tensions. It is a fact that often different generations experienced very different forms of socialization and these authors pointed to a certain inevitability of culture wars emerging between the young and the old. Such question prefigures later concerns in the wider society about the question of youth and how they could be seen as harbingers of disorder, change and conflict (Cohen,1997).

Through the latter part of the twentieth century in affluent societies, it has been evident that the phase of youth extends from the teenage years beginning at the age of fourteen and fifteen to the early twenties and beyond. This is due to the fact that many youth spend longer periods of time for education and training and delay their entry into full-time work, family and household formation. For many, such delays are a result of unemployment, poor quality of work and social exclusion. In recent times commentators have spoken about a 'boomerang generation' of young people who are in their twenties and thirties who after considerable efforts trying to secure independent homes and work return home because of the high cost of housing. It is in this context that the state has considerable power over the youth phase as it has influence over education and labour markets. Governments in affluent societies have increasingly stressed upon the upskilling of its citizens in order to create 'knowledge economies' where workers would be able to acquire higher credentials in terms of education (Lauder et.al., 2007). Thus, over a period of fifty years we have witnessed the significant extension of compulsory education from school leaving age of thirteen to fourteen to one where most young people are engaged full-time in education until the age of eighteen and a majority until twenty-one years of age. Operating with definitions of youth transitions and youth identities, youth studies professionals have made efforts to offer an analytic map of the phase of youth. The phase transition is usually understood as encompassing multiple routes into adulthood in relation to key aspects of young people's lives such as education and employment, intimate relationships and friendships, housing and leisure. There are some who argue that education and employment are fundamental which helps in structuring the other pathways of transition (Roberts, 2003). In affluent societies, in recent years, some commentators discuss the breakdown of heavily structured and predictable transitions along classed and gendered lines and hence have moved away from using metaphors such as careers and routes (Banks et.al., 1992) that denote transitions to metaphors that denote to greater fluidity such as navigation and niches (Evans and Furlong, 1997). Researchers who have developed images of youth transitions have imaged them as socially constructed since they reflect the strong influences of culturally and historically specific events such as deindustrialization of many European countries during the 1980s and 1990s and the economic growth and cultural transformation of many cities in developing societies during the first decade of the twenty first century (Farrar, 2002).

Researchers in youth studies have also developed models of young people's social identities that help us in understanding how young people are defined. During the 1970s youth were often understood in relation to crude notions of their structural class, race and gender positioning in the wider society (Mungham and Pearson, 1976). However more recently, concepts of youth identities that define young people as living through multi-faceted processual notions of the self where individuals undertake identity work and identity performances creating hybridized identities have superseded constructs that pointed towards youth belonging to a structural class and gender positioning (Bennett, 1999; 2005). Such developments in terms of identity theory have been framed by the growing influence of media based interactions and diasporic, migratory experiences of many young people in recent times. Writers still acknowledge that economics, social relationships and cultural formations have a bearing on the framing of youth identities.

However, there are many who emphasize on the loosening of conditioning processes that enable young people to have more space and opportunity to create their identities across what were once rigid and impermeable boundaries (Pysnakova and Miles, 2010). The hybrid selves of young people can also be understood in the context of the greater sensitivity of reflexive processes, so-called internal conversations and self-monitoring, in which we all participate and which constitutes our daily lives. Young people today if seen in this context, are more mindful of their identity vis-à-vis previous generations and are careful of their aims in life and their self-development (Beck, 1992; Giddens, 1991). Thus, any definition of youth today should be sensitive to historical processes such as individualization and detraditionalization that have had an impact on how young people conceive who they are and how they live their lives.

Although definitions of youth vary from country to country, it is the definition of the United Nations (UN) that defines youth as those persons between the ages of 15 and 24 years without prejudice to other definitions by member states. According to the International Labour Organization (ILO) Convention 138, youth begins when a person reaches the age of 15. The Commonwealth Youth Programme defines youth as those in the age group of 15 to 29 years. Many countries interpret the achievement of the age majority, the age at which a person is given equal treatment under law as the entry from the state of youth to adulthood. However, the operational definition and nuances of the term youth often varies from country to country depending on the specific socio-cultural, institutional, economic and political factors (Johal, Kaur, Begra, Manchanda, 2012). Eisenstadt (1956) defined youth as the period of transition from childhood to full adult status with full membership in the society. As a stage of human development, youth is a phase of high expectations, high risk taking and great enthusiasm and therefore, is a strong force to reckon within society. They can be mobilized for physical target achievements and for psychological purposes by utilizing their capacity for sacrifice, courage, endurance and initiative. Youth is also a period of training and acquisition of skills. Rosenmayr (1972) identifies five conceptual approaches in defining youth: (i) youth as a phase in the individual life-cycle – the psychological and biological growth; (ii) youth as a social subset: forms of behaviour in a roughly determined age range; (iii) youth as an incomplete status: transition between childhood and adulthood; (iv) youth as a socially structured generation-unit: experiencing certain common conditions and generating common activities; (v) youth as an ideal value concept: idealism, alertness, traits called youthfulness (Youth Development Report, 2010).

In the year 1978, Pierre Bourdieu teased with the idea – 'la "jeunesse" n'est qu'un mot' ('youth is just a word'), however, words cannot be taken merely as words and even artificial constructs have social meanings that have real effects. Therefore, any discussion in terms of youth practice, youth policy or youth research raises difficult and challenging questions about its meanings and its conception, its social and historical construction as well as with social and political implications for our understanding of young people's lives.

While scholars, commentators and policy makers have examined key actors in the peacebuilding processes, one important actor – youth – seems to be, although in very limited terms, recognized in recent years as potential agents in the peacebuilding processes, while it is interesting to know how youth can be engaged in peacebuilding processes, it is also imperative to examine why youth engage in conflict and subsequently those theories which substantiate them. This volume thus, has a focus on the conflict in north east India where youth participation in it and how their engagement in the peacebuilding processes can bring sustainable peace in the region. The argument that this book posits is youth can be engaged, notwithstanding the challenges and hurdles, positively in building peace in the region. The chapters that follow makes a critical effort to help the reader to have a clear understanding on conflict and peacebuilding, youth participation in conflict, youth engagement in peacebuilding and how youth and their potential can be utilized to build peace in north east India can be made.

REFERENCES

Allen, S. (1968) 'Some Theoretical Problems in the Study of Youth' The Sociological Review, 16(3): 319-331.

Aries, P. (1965) 'Centuries of Childhood: A Social History of Family Life', New York: Vintage Books.

Banks, M., Bates, I., Brekwell, G., Emler, N., Jameison, L., and Roberts, K. (1992) 'Careers and Identities', Milton Keynes: Open University Press.

Beck, U. (1992) 'Risk Society: Towards a New Modernity', London: Sage Publications.

Bennett, A. (2005) 'In Defence of Neo-Tribes: A Response to Blackman and Hesmondhalgh', Journal of Youth Studies, 8(2): 255-259.

Boyden, J (1990) 'Childhood and the Policy Makers: A Comparative Perspective on the Globalization of Childhood', in James, Allison and Prout, Alan (eds.) Contemporary Issues in the Sociological Study of Childhood, Philadelphia: Falmer Press.

Cohen, P. (1997) 'Rethinking the Youth Question: Education, Labour and Cultural Studies', London: Macmillan.

Coles, B. (1995) 'Youth and Social Policy', London: University College London.

Connell, R.W. (1987) 'Gender and Power', California: Stanford University Press.

Davies, B. (2004) 'Curriculum in Youth Work: An Old Debate in New Clothes?', Youth and Policy, 85(1): 87-98.

Davies, B. (2006) 'the Place of Doubt in Youth Work: A Personal Journey', Youth and Policy 92(1): 69-80.

Edwards, A.R. (1983) 'Sex Roles: A Problem for Sociology and for Women', Journal of Sociology, 19(3): 385-412.

Eisenstadt, S.N. (1956) 'From Generation to Generation: Age Groups and Social Structure', New York: The Free Press.

Erickson, E. (1968) 'Identity: Youth and Crisis', New York: Norton Press.

Evans,K. and Furlong, A. (1997) Metaphors of Youth Transitions: Niches, Pathways, Trajectories or Navigations', in Bynner and Furlong, A. (eds.) Youth Citizenship and Social Change in a European Context, Aldershot: Ashgate Publications.

Farrar, S. and Lowe, J. (1978) 'Sex Role Theory: Political Cul-de-Sac', Refractory Girl, 6(1): 14-16.

Gidden, A. (1991) 'Modernity an Self Identity: Self and Society in the Late Modern Age', Cambridge: Polity Press.

Gillis, J (1994) 'Youth and History: Tradition and Change in European Age Relations 1770-Present', New York Academic Press.

Hall, G.S. (1904) 'Adolescence: Its Psychology and its Relation to Physiology, Anthropology, Sociology, Sex, Crime, Religion and Education', New York: Appleton and Co.

Head, B.W. (2011) 'Why not Ask Them? Mapping and Promoting Youth Particiaption', Children and Youth Services Review, 33(4): 541-547.

Irwin, S. (1995) 'Social Change and the Transition from Youth to Adulthood', London: UCL Press,

Johal, R.K., Kaur, J., Begra, S., and Manchanda, P.K. (2012) 'Situation Analysis on Youth and Local Governance', Chandigarh: Commonwealth Youth Programme, Asia Centre.

Jones, G. (1988) 'Integrating Process and Structure in the Concept of Youth: A Case for Secondary Analysis', The Sociological Review, 36(4):706-732.

Jones, G, and Wallace, C. (1992) 'Youth, Family and Citizenship', Buckingham: Open University Press.

Lauder, H., Brown,P., Dillabough, J. and Haley, A.H. (eds.) (2007) 'Education, Globalisation and Social Change', Oxford: Oxford University Press.

Liebau, E. and Chisholm (1993) 'Youth, Social Change and Education: Issues and Problems', Journal of Education Policy, 8(1):3-8.

Mannheim, K. (1952) 'The Problem of Generations', in Mannheim, K. Essays on the Sociology of Knowledge, London: Routledge and Kegan Paul.

Marshall, T.H. and Bottomore, T. (1992) 'Citizenship and Social Class', London: Pluto Press.

Mungham, G. and Parsons, G. (eds.) (1978) 'Working Class Youth Culture', London: Routledge and Kegan Paul.

Parsons, T. (1942) 'Age and Sex in the Social Structure of the United States', American Sociological Review, 7(1): 604-616.

Pysnakova, M. and Miles, S. (2010) 'The Post-Revolutionary Consumer Generation: Mainstream Youth and the Paradox of Choice in the Czech Republic', Journal of Youth Studies, 13(5): 533-547.

Rajiv Gandhi National Institute of Youth Development (2012) 'Youth Development Report 2010', Sriperumbudur: Rajiv Gandhi National Institute of Youth Development.

Wallace, C and Cross, M. (eds.) (1990) 'Youth in Transition: The Sociology of Youth and Youth Policy', London: Falmer Press.

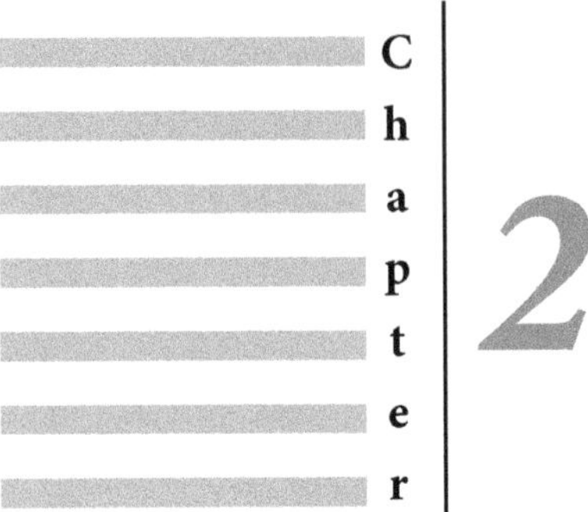

Understanding Conflict and Peace Building

Behind every analysis of violent conflict is a set of assumptions. Assumptions about what moves human action and how to study it and how the interests, instincts, structures or choices that explain why and how people resort to violence. These assumptions are usually very basic and fundamentally subjective. Assumptions form the base of academic theories of conflict. Indirectly they also inform the ways policy makers and politicians read a conflict. Their interpretation of a conflict determines – to a certain extent – what sort of intervention policy makers and politicians design. If a conflict is understood as stemming from ancient hatreds between ethnic groups, there is little outsiders can do: for as soon as third parties pull out, old animosities will flare up again and violence will be resumed. Lack of grounded and critical analysis of violence and war results in misreading and framing inaccurate strategies and interventions, with at times dramatic consequences. This calls for a defined specialist field of study which would provide analytical frames, research methodologies and skills to explain and understand contemporary violent conflict. Why and how wars happen? Why are people prepared to kill and die in the name of an ethnic or religious group? How are people mobilized to join in? how can neighbors turn into enemies? And more fundamentally: what is the role of identity deprivation, structural change, rationality and discourse? What is the social media and the internet on repertoires of contention, and what is the connection between the global spread of neo-liberalism and local forms of violence? There has been sustained debate in academic disciplines on the causes, dynamics and consequences of violent conflict. However, despite the wealth of material on violence and war, there remains much to understand and learn from earlier insights and theories (Demmers: 2012).

Conflicts occur at various levels ranging from international levels of war to interpersonal level. For our present purposes, however, a dichotomous distinction between conflicts within and between states seems adequate, albeit only with the addition of their combination in internationalized inter-state conflicts, here labeled transnational conflicts. It further makes sense to distinguish between violent and non-violent conflicts and to acknowledge that neither the borders separating the various levels nor those between violent and non-violent conflicts are entirely clear or impermeable (Moller:2003).

In recent years/decades the nature of large-scale violent conflicts has fundamentally changed from an era of wars to one that is characterized by complex political emergencies (CPES). A number of conceptual shifts have occurred in the attempt to better understand the nature of these multiple small wars. Classic analytical frameworks focusing on the relationships between states, military capacities and strategies and international political economy are being put aside for more eclectic frameworks. These draw heavily on social and cultural theory, blend different theoretical elements together to analyze different situations, relate conflict to development and point to the inherent unpredictability of conflict processes and outcomes (Goodhand and Hulme: 1999). The years after 1989 saw more military operations in more parts of Europe, Asia and Africa than anyone could remember, since it was often unclear who was fighting whom and why. 'These activities did not fit under any of the classic headings of war – international or civil... the century ended in global disorder whose nature was unclear' (Hobsbawm: 1994; Goodhand and Hulme: 1999).

It is clear, that the shift in patterns of violent conflict from wars between states to conflict within states, which began around the middle of the twentieth century, has not been reversed by the end of the cold war. Of the eighty two armed conflicts between 1989 and 1992 only three were between states. More than half of these eighty two conflicts had been underway for atleast a decade (UNDP:1994). While death and disablement are a common feature of both classic wars and contemporary CPES (Burton: 1994), there has been a horrifying shift in the distribution of suffering and nowadays around ninety percent of casualties are inflicted on civilians. Changes in the nature of violent conflict and the contexts within which it is set have required changes in the concepts that are used to aid the understanding of contemporary conflict. Analyses that are focused on the relationships between states, on military capacities and strategies, on predicting who would win and who would lose and an international political economy have increasingly been replaced by social and cultural analyses that recognizes the complexity and contingency and the question of feasibility of prediction. This chapter reviews the contemporary approaches to the understanding of conflict and peacebuilding and highlights the continuing importance of relating the discourse about conflict and peace to the discourse that is around social and economic development.

Since the end of World War II, a total of 240 armed conflicts have been active in 151 locations around the world. In the year 2008, thirty six conflicts were active in twenty six locations worldwide. This is one more than what was recorded in the year 2007. While the number of active conflicts has not seen any drastic changes from one year to the next, it has gone up by seven or nearly one quarter since 2003, the year with the lowest number of conflicts since the 1970s. Yet, the number of conflicts remains at only two thirds of the peak recorded in 1992. The biggest increase occurred in Africa from nine in 2003 and seven in 2005 to 12 in 2008. In the years 2006 and 2007, conflicts re-erupted in the Democratic Republic of Congo (DRC) Somalia, Niger and Mali and an entirely new conflict broke out in western DRC (Harbom and Wallesteen: 2009).

Until the end of the cold war, the conventional wisdom in the world was that ethnicity and nationalism were outdated concepts and largely resolved problems. On both sides of the cold war, the trend seemed to indicate that the world was moving toward internationalism rather that nationalism. As a result of the threat of nuclear warfare, great emphasis on democracy and human rights, economic interdependence and gradual acceptance of universal ideologies, it became fashionable to speak of the demise of ethnic and nationalist movements. Despite expectations to the contrary, however, a fresh cycle of ethno-political movements have re-emerged recently in eastern Europe, central Asia, Africa and South Asia and many other parts of the world. Infact with the end of the cold war, which has clearly increased international cooperation while decreasing the possibilities of inter-state wars, the main threat to peace does not come from major inter-state confrontations anymore, but from another source – intra-state conflicts or conflicts that occur within the borders of states. These conflicts have replaced the cold war's ideological clashes as the principal sources of current conflict. The fact that internal conflicts occur within the borders of states made major international actors reluctant to intervene as well, either for legal concerns or concern to avoid probable loses. Thus unless conflicts really escalate, the international community has preferred not to get involved in intra-state conflicts. But such conflicts are as serious, costly and intense as any in the past. These conflicts somehow need to be managed and reduced, or else international peace and security will not be in a stable situation. Even if intra-state conflicts appear to be local, they can quickly gain an international dimension due to global interdependence and to various international support. Infact, when external parties provide political, economic, military assistance, asylum or the bases for actors involved in local struggles, these conflicts inevitably assume an international dimension. Undoubtedly, the effective management of intra-state conflicts by domestic and international authorities presupposes an understanding of their nature and causes (Yilmaz:2007). Others in contrast have noted that the post World War II era has witnessed a great deal of warfare. Most brutally tens of millions have died in combat in the modern era (Small and Singer: 1982) and even more in the acts of genocide, politicide and democide that often accompany war and are exacerbated by it (Harff and Gurr: 1988). There are those who claim that both international and civil wars have made a comeback (Clodfather: 1992). Other scholars have focused more specifically on the emergence of possibly new types of war. For instance, Kal Holsti has argued that we are seeing the emergence of new wars of the third kind and others decry the proliferation of ethnic wars (Lake and Rothchild: 1998). For all these reasons and more, it is vitally important to know whether war is ubiquitous as some say, declining as fast as others assert, or fluctuating across time and place. Surprisingly, though hundreds of books and articles have tangentially addressed this question in recent years, too few have examined and answered the basic question of whether the data indicate that the total amount of war in the world is increasing or decreasing or remaining the same. Research endeavors have tended to focus on only one type of war and many of the contradictory conclusions are primarily directed by the type of war being studied (inter-state or civil). Obviously part of this dichotomy is linked to cleavages in the discipline, with international politics scholars primarily concerned with wars between states and scholars in the field of comparative politics have their attention focused on internal developments that include revolutions and civil wars (Gurr: 1970; Midlarsky: 1988). The lack of a dialogue between these two realms has limited the accumulation of knowledge and has circumscribed our understanding of war. More particularly, following the end of the cold war, a narrow disciplinary focus exclusively on inter-state wars in no longer justifiable (Eberwin and Chojnacki: 2001).

Wars between states have declined sharply in number since the end of World War II along with the founding of the United Nations Organization. This is primarily the result of a generalized attempt to avoid further conflicts. Thus, the UN Charter vows to make every effort to save succeeding generations from the scourge of war by identifying and addressing threats to peace, breaches of peace, use of force or acts of aggression. War amongst people which is referred as intra-state war makes a paradigm shift in the approach to understanding conflicts. In such situation, not only are the participants different, but also the environment. Intra-state wars mainly oppose state to feudal, often warlike, organizations which seek to usurp state prerogatives. To simplify, these feudal organizations can be terrorist groups, guerillas, organized crime syndicates or a combination of these types of organizations. The majority of these are under the domination of more powerful groups whose goal is the conquest of local or national sovereignty – for example, the Revolutionary Armed Forces of Colombia (RAFC), Afghan Taliban, the Lebanese Hezbollah or the Liberation Tigers of Tamil Eelam (LTTE). They try to acquire the attributes of state, by elections, violence, corruption, ideological persuasion or political compromise. They hold sway over a territory and its population and setup political, military, social and legal institutions. They use a comprehensive approach to totalitarian methods to achieve their aim. It is not in all cases that the smaller ones want to overthrow the state, but they look to substitute themselves to it locally by preserving a less rigid form of governance. They acquire the privileges of the state without its structure and its duties (Dosse: 2010).

That which sets apart the first world from the third world appears significantly in terms of the frequency internal conflicts and political violence occurring in the latter. Many third world countries experience a higher rate of internal conflicts and political violence, yet this may not be true in all cases vis-à-vis their first world counterparts. Here the dichotomy between conflict stricken and non-occurrence of internal conflict and political violence needs to be addressed by analyzing the determinants of internal conflicts and identifying those conditions that dichotomize between occurrence and non-occurrence of internal conflicts in third world countries. The attacks on the United States on September 11 and the subsequent wars in Afghanistan and Iraq have overshadowed crises and internal wars prevalent in other parts of the world particularly in the third worlds. Crises in other parts of the third world have become forgotten crises when indeed they are still active and significant (United Nations: 2004).

Conflict stricken countries in the third world in most cases struggle with poverty, civil wars, refugees, political violence and instability, shortage of food, drought, AIDS, as well as economic devastation. Internal conflicts or crises appears or manifest in many forms such as ethnic, religious conflicts, riots and attempts to overthrow governments forcefully. Secessionist or independence movements, or civil or political violence may arise due to a number of reasons such as racial, religious, cultural, ideological and economic factors as well as political and social structures. In recent decades there has been a phenomenal increase in the usage of terms such as ethnic conflict, ethnocentrism, ethno nationalism, and ethnic cleansing where the prefix 'ethnic' has been more widely used since the end of the cold war. These terms also portray internal conflicts as being based on ethnic-racial lines within a state. In third world countries, ethnic-racial determinists of internal conflicts arguments are based on the distinctiveness of each of the diverse groups that exist within a state as the very source of internal conflict and political violence (Hall: 2004). Religious determinists on the other hand emphasize that religion or religious differences are the source of internal conflict. Religion is the source of identity and if their system of belief and identity, if compromised, serve as key factors in the emergence of civil wars and political violence. Some also argue that where religion is the source of conflict, it could spill over into ethnic conflicts as well (Jurgensmeyer: 1993).

Throughout history there has been incidents of ethnic discrimination, exploitation, and conflict that have been captured and broadcast in news and television screens which has caught the attention of many in the world. Many countries in the world have been witnessing conflicts between ethnic groups that have sometimes been on a horrific scale. Less visible and news worthy and much more pervasive are non-violent conflicts that can appear in multiple forms. Ethnic groups in some countries compete through parties that are overtly ethnic and vie for power. In most cases the dominant ethnic group exploits and discriminates other groups in the state. Esman puts in succinctly that 'when and ethnic group gains control of the state, important economic assets are soon transferred to the members of that community (Esman: 1994). By suggesting that ethnicity helps enforcing the dominant ethnic group's monopoly of the state's assets, we hope to explain these observations. Yet and crucially, ethnic conflicts are not universal in nature and in ethnically heterogeneous societies that can be found in many countries, ethnic groups co-exist peacefully. Nor can we assume that in ethnically heterogeneous societies that there will be fairly harmonious ethnic relations before or after periods of conflict. In this context, it would be good to ask why do some countries with multi ethnic groupings experience conflict and others don't? Why do ethnic conflicts wax and wane over time in the same country?

Conflicts that are ethnic in nature are of course a recurrent phenomenon. The international environment often plays a part in its emergence and remission. Often overshadowed by international warfare and masked by wartime alliances, ethnic allegiances are usually revived by the wartime experience or reemerge soon after the war as it did after the first and the second World Wars. In their periodic reemergence, ethnic sentiments have been supported by the widespread diffusion of the doctrine of national self-determination (Connor: 1967). This doctrine that can be traced back to the eighteenth century conceptions of sovereignty, flowered with the burst of nationalism in the nineteenth century Europe and the emergence of states like Germany and Italy out of more parochial units. On the other hand, the twentieth century version, entails the dismemberment of empires and large states in favor of smaller units that began following the end of the first World War which along with the Wilsonian espousal of self-determination helped in the remaking of the map of Europe (Horowitz: 1985).

This process was repeated following the end of the second World War which saw the end of the colonial rule in Asia and Africa. The process of decolonization set in motion a chain reaction the ultimate impact of which are still being felt today. The movements that sought independence from colonial powers were not always wholly representative of all the ethnic groups within their respective territories. Some groups that were not so well represented, with varying degrees of success, attempted to slow down the march towards independence in order to gain special concessions of even the creation of a separate state. However, with some exceptions ethnic differences tended to be muted until the process of independence was completed. Following independence, however, the context and issues changed. It was no longer the issue of colonial domination. Self-determination had been implemented only to the level of pre-existing colonial boundaries. It was within these boundaries that the question of to whom did the new states belong. While some groups moved to succeed to the power of the former colonialists, others were heard to claim that self-determinations were still incomplete as they had not achieved their own independence. Certain worldwide ideological and institutional currents have also reinforced the growth of ethnic conflicts. The spread of norms of equality had made ethnic subordination illegitimate and spurred ethnic groups everywhere to compare their standing vis-à-vis groups that were in close proximity with them. The simultaneous spread of value of achievement has cast in doubt the work of such groups whose competitive performance seems deficient by such standards. The state system that first grew out of European feudalism and now in the post-colonial period seems to have had its influence over all countries over the world provides the framework under which ethnic conflicts occur. In consequence to all these developments, ethnic conflict possesses the elements of universality and uniformity that were not present at earlier times. The ubiquitous character of ethnic conflict opens opportunities for groups and movements to become part of a broad and respectable current, learning from each other and in doing so becoming similar in their claims and aspirations. The profusion of ethnic claims is infact expressed in a distinctly parsimonious common rhetoric. Its terminology is the language of competition and equality, a remarkably individualistic idiom for claims that are advanced on a collective basis for comparative analysis, for ethnic conflict has common features (Horowitz: 1985).

Drawing on international relations theory makes sense for several reasons. International relations theory is primarily concerned with issues of war and peace. Which one has to be cautious and avoid straight forward translation of findings from the realm of inter-state relations to those of inter-ethnic/inter-group relations, it is equally important to bear in mind that some of the units of analysis are of course the same, not least if one subscribes, as we do, to the idea that it is after all individual-leaders as well as followers who have choices to make about war and peace or conflict or co-existence. Even though theories of international relations are concerned with the role and behavior of states in the international arena, they nevertheless start by making fundamental assumptions about human nature. Realism and liberalism both consider human beings as self-centered and rational actors concerned with their own survival. In an anarchical world – the Hobbesian state of nature – this translates readily into a complete reliance on self-help, that is, acquire as much power as you possibly can in order to defeat any threat to your survival. While proponents of the two traditions and their various sub-schools differ is the extent to which this is not only the natural state of affairs, but one that exists in perpetuity. Realists are generally pessimistic about human nature, while liberalists are optimistic about human beings being capable of learning from experience (Cordell and Wolff: 2010).

The way the international community understands peace and security is shaped in two way eversince the end of the cold war. First, the range of political actors that could be engaged in conflict has expanded significantly which includes a number of non-state entities. The idea of security is no longer narrowly conceived in terms of military threats that could arise from aggressor nations. In developing countries, state failure and civil wars represent some of the greatest risks to global peace. Countries that are war torn have become recruiting grounds for international terrorist networks, and havens for organized crime and drug traffickers and as a result, tens of millions have crossed borders as refugees and created new tensions in host communities. Instability that has been caused by cross-border incursions by rebel groups could have a ripple effect causing disruptions in trade, tourism and international investments. Second the potential causes of insecurity have also increased and diversified considerably. While political and military issues remain critical, conceptions of conflict and security have broadened which now include economic and social threats such as poverty, infectious diseases and environmental degradation that are today seen as significant factors contributing to the emergence of conflict. This new understanding of the contemporary challenges to peace is now being reflected in high-level policy debates and statements. The 2004 report of the UN Secretary-General High-Level Panel on Threats, Challenges and Change highlighted the fundamental relationship between the environment, security and social and economic development in the pursuit of global peace in the 21st century (UNGA: 2004). While a historical debate at the UN Security Council in June 2007 concluded that poor management of high value resources constituted a threat to peace (UN Security Council 2005), more recently, UN Secretary General Ban Ki-Moon confirmed that the building blocks of peace and security for all peoples are economic and social security anchored in sustainable development because they allow us to address all issues relating to poverty, climate, environment and political stability – as parts of a whole. The potential for conflict to be ignited by the environmental impacts of climate change is also attracting international interest in this topic. In the year 2009, the European Union in a high-level brief stated that climate change for instance could be a threat multiplier which might exacerbate existing trends, tensions and instability posing both political and security risks (UNEP: 2009).

The agenda for peace is one of the cornerstones in international debates on how to build peace after violent conflicts. Peacebuilding right from the beginning was not regarded as a concept that would seek to transform societies in or emerging from conflict, but to maintain stability. If conflict is caused, enabled or reproduced by certain particular social structures and institutions which favor a dominant group, we cannot hope to or remove or alleviate those causes without altering those structures. Then peacebuilding becomes another aspect of a system which only seeks stability within the confines of that system which already made war possible (Featherston: 2000).

The framing of peace and ways of building it led to a preferred set of methods and methodologies. As elsewhere in development or qualitative research such as that of Paul Collier, for the World Bank (Collier et.al: 2003) – gained prominence offering a powerful instrument to legitimize interventions by aid organizations. However, the deployment of such approaches to research not only served to erase the particularity to places and experiences through its inevitable generalization, but also had further costs (Denskus: 2007).

In order to better identify what institutionalizes peace after war, scholars, practitioners international and regional organizations have attempted to point out what those critical ingredients are that would further the goal of peace. If the success of peacebuilding is measured against how well it has institutionalized peace, the picture is very mixed. Nearly every country receiving assistance slide back into conflict within five years and seventy-two percent of peace building operations place authoritarian regimes. Success if measured in terms of institutionalization of the concept of peacebuilding, then it appears to be an unquestionable success. Peacebuilding is a concept that is being used by an impressive number of organizations that work for the cause of preventing or ending deadly conflicts and use it to frame and organize post-conflict activities. The fact that weak states often experiencing civil wars pose a major threat to international stability has now been joined with the long term concern and demand for building peace. In this context the most important sign towards peacebuilding agenda appeared in the 2005 World Summit at the UN where the proposal by UN Secretary General to establish a Peace Building Commission, support office and fund were endorsed. This has helped in institutionalizing the peacebuilding agenda at the highest levels and the motivation for others to join the peacebuilding bandwagon. Peacebuilding although is generally defined as external interventions that are designed to prevent the emergence or relapse into armed conflicts, there remains critical differences in terms of its conceptualization and operationalization among actors (Barnett, Kim, O'Donnell and Sitea: 2000).

Sustained peace is not a naturally occurring phenomenon - it must be consciously and continuously constructed and renewed. This is the central premise of what the international community and the scholarship that has grown up around it call peacebuilding. It applies not only to the relations between states, but also in terms of the contentious politics that exist within them. States that have recently suffered violent conflicts clearly face daunting challenges than those with no recent history of civil war. Chronic recidivists such as Afghanistan and the Democratic Republic of Congo (DRC) are just the visible face of a larger problem. While it may be hard to support as a general proposition, the widely propagated assertion that half of all post-conflict countries return to war within five years, the recrudescence of violence is an ever looming threat. Most post conflict societies that have avoided returning to the conflict category have suffered extremely high rates of violent crime with people forced to endure levels of insecurity exceeding those in many officially designated zones of conflict. The idea of peacebuilding has risen to prominence in all three sectors of the UN work – international security, sustainable development and human rights. Different shades of the meaning are conveyed by different users of this capacious term, but at its core, peacebuilding is about preventing the outbreak or recurrence of widespread and systematic violence. Peacebuilding is generally associated with efforts to move beyond immediate technical problems such as lack of physical administrative or economic infrastructure (all of which are acute in post-conflict settings), to a more encompassing, partly political approach that engages with the primary actors involved in conflict and addresses its causes, particularly those stemming from developmental deficits or the basic structure of the political settlement. Peacebuilding, undertaken by both national and international authorities or non-governmental organizations, is related to but distinct from peacekeeping – a term better known to the public, but one which does not like peacebuilding appear in the UN Charter (Jenkins: 2013).

The concept of peacebuilding emerged in response to the growing concern that traditional development approaches might not be adequate or appropriate in situations characterized by insecurity, heightened risks of conflict and weak state capacity and legitimacy. In general terms, peacebuilding is about ending or preventing violent conflict and supporting sustainable peace. A range of measures are involved in peacebuilding that are targeted at reducing the risk of relapsing in conflict. This is done by strengthening national capabilities at all levels to manage conflict and to lay the foundation for development and sustainable peace. Key peacebuilding objectives include inter-alia, preventing countries from relapsing into violent conflict, establishing structures and incentives for peaceful mitigation of conflicts, incentivizing elite commitment to peace processes while laying the groundwork for those processes to be more inclusive overtime, establishing a framework of political security and economic transition, jumpstarting recovery, demonstrating peace dividends by meeting the urgent needs of the populations (OECD: 2010).

While the term peacebuilding is relatively new, external assistance for post-war rebuilding goes back to the reconstruction of post-World War II Europe and Japan. The promise of new peacebuilding agenda was that the international community would intervene collectively as a third party in order to help resolve violent conflicts and civil wars without affecting the national interests of individual states. The impetus for peacebuilding came from multiple sources with its strongest expression appearing at the UN. The UN throughout the 1990s provided the rationale and the operational principles for post-conflict peacebuilding as one of a series of tools at the UN's disposal alongside preventing diplomacy, peacemaking and peacekeeping. In order that peacekeeping and peacemaking to be truly successful, most come to identify support structures that would have the capacity to consolidate peace and advance a sense of confidence and well-being among people. Agreements that end civil strife in most cases include disarming the previously warring parties and the restoration of order, the custody and possible destruction of weapons, repatriating refugees, advisory and training for security personnel who monitor elections, enhancing efforts to protect and promote human rights, strengthening and reforming governmental institutions and promoting formal and informal processes of political participations. The 1995 supplement to 'An Agenda for Peace' for example, noted the linkages between conflict prevention and peacebuilding through demilitarization, control of small arms, institutional reforms, improved judicial and police systems, monitoring of human rights, electoral reform and social and economic development can act as valuable instruments in preventing conflict and also in healing wounds that had occurred due to conflict. It also acknowledged that implementing peacebuilding could be complicated – requiring integrated action and delicate dealings between the UN and the parties to the conflict in respect of which peacebuilding activities are to be undertaken. The international approach to peacebuilding and conflict prevention is grounded in the concept of liberal peace, which derives from a long tradition of western liberal theory and practice. The liberal peace thesis views political and economic liberalization as effective antidotes to violent conflicts. Thus, promotion of human rights, democracy, elections, constitutionalism, rule of law, property rights, good governance and neo-liberal economics have become part of the peacebuilding strategy (Tschirgi: 2004).

Building peace in countries that are emerging from conflict is a huge and complex undertaking. It involves a myriad of different players and stakeholders beginning with citizens of those countries where peace is being built. It cannot be said to be purely political, security or a developmental process but one that brings together security, political, economic, social and human rights elements in a coherent and integrated way. In most cases, the international community looks to the UN to lead the international response and coordinate the actors involved. Given the regional dimensions of many contemporary conflicts, regional organizations are increasingly at the forefront of peace processes, including in mediating and guaranteeing peace agreements and monitoring their implementation. Regional actors may sometimes be better equipped that global actors to adapt and apply universal principles in a local context.

Today the world is home to the largest population of youth that has ever been recorded. About 1.2 billion youth make up for eighteen percent of the world's population. Being caught between child and adulthood, voices of youth are distinctively underrepresented on issues that concern them – including issues of peace and conflict. A UNSC resolution which specifically focuses on youth participation in peacebuilding has, to date never been adopted. It is exactly the absence of such a resolution that contributes to the unparalleled underrepresentation of youth in peacebuilding. Armed conflict is one of the most critical challenges that young people are facing today. Youth are the main victims of conflicts, not only because of direct violence perpetrated against them but also because of their unique vulnerability to both voluntary and involuntary military recruitment or induction into armed rebel groups. These assertions have led to a dichotomous view point of youth as either casual or recipient agents. I however, argue for a third viewpoint that is one in which youth are recognized as agents for peace.

REFERENCES

Barnett, M., Kam, H., O'Donnell and Sitea (2007) 'Peacebuilding: What's in a Name?', Global Governance, 13(1):35-58.

Burton, J (1994) 'The Upsurge in Interest in the Relief-Development Continuum: What Does it Mean', Relief and Rehabilitation Network, Newsletter, September 1994.

Clodfetter, M. (1992) 'Warfare and Armed Conflicts: A Statistical Reference to Casualty and Other Figures 1618-1991', North Carolina: Jefferson McFarland.

Conca, K., and Dabelko, G.D. (2018) 'Green Planet Blues: Critical Perspectives on Global Environmental Politics', London: Routledge.

Connor W. (1967) 'Self-Determination: The New Phase', World Politics, 20(1): 30-53.

Cordell, K. and Wolff, S (2010) 'Ethnic Conflict', Cambridge: Polity Press.

Demmers, J. (2012) 'Theories of Violent Conflict: An Introduction', New York: Routledge.

Denskus, T. (2007) 'Peacebuilding Does not Build Peace', Development in Practice, 17(5): 656-662.

Dosse, S. (2010) 'The Rise of Intra-State Wars: New Threats and New Methods', Small Wars Journal, www.smallwarsjournal.com. Accessed: 19.02.2019.

Eberwein, W. and Chojnacki, S. (2001) 'Scientific Necessity and Political Utility: A Comparison of Data on Violent Conflicts', Discussion Paper P01-304, Berlin: Center for Social Science Research.

Esman, M. (1994) 'Ethnic Politics', New York: Cornell University Press.

Fetherston, B. (2000) 'From Conflict Resolution to Transformative Peacebuilding: Reflections from Croatia', Working Paper 4. Bradford: Center for Conflict Resolution.

Goodhand, J and Hulme, D. (1999) 'From Wars to Complex Political Emergencies: Understanding Conflict and Peacebuilding in the New World Disorder', Third World Quarterly, 28(1): 13-26.

Gurr, T. (1970) 'Why Men Rebel', Princeton: Princeton University Press.

Hajer, M.A and Wagenaar (2003) 'Deliberative Policy Analysis: Understanding Governance in the Network Society', Cambridge: Cambridge University Press.

Hall, T.D. (2004) 'Ethnic Conflicts as a Global Social Problem in George Ritzer (ed) 'Handbook of Social Problems: A Comparative International Perspective', Thousand Oaks; Sage Publications.

Harbom, L. and Wallensteen, P. (2009) 'Armed Conflicts 1946-2008', Journal of Peace Research, 46(4): 577-587.

Harff, B. and Gurr, T. (1988) 'Toward Empirical Theory of Genocide and Politicides: Identification and Measurement of Cases Since 1945', International Studies Quarterly, 32(3): 359-371.

Hobsbawm, E. (1994) 'Age of Extremes: The Short Twentieth Century 1914-1991', London: Abacus.

Horowitz, D.L. (1985) 'Discourses on Violence: Conflict Analysis Reconsidered', Manchester: Manchester University Press.

Jenskins, R. (2013) 'Peacebuilding and State Building Priorities and Challenges: A Synthesis of Findings from Seven Multi-Stakeholder Consultations', Dili, Timor-Leste: Organization for Economic Cooperation and Development.

Jurgensmeyer, M. (1993) 'The New Cold War? Religious Nationalism Confronts the Secular State', Berkeley: University of California Press.

Lake, D.A. and Rothchild, D. (1998) (eds.) 'The International Spread of Ethnic Conflict', Princeton: Princeton University Press.

Midlarski, M. (1988) 'Rulers and Ruled: Patterned Inequality and the Onset of Mass Political Violence', American Political Science Review, 82(2): 491-509.

Moller, B. (2003) 'Conflict Theory', Working Paper Series, 122, Denmark – Aalbory: DIR and Institute for History, International and Social Studies.

Small, M. and Singer, J.D. (1982) 'Resort to Arms: International and Civil War, 1916-1980', California: Sage Publications.

Tschirgi, N. (2004) 'Post-Conflict Peacebuilding Revisited: Achievements, Limitations, Challenges', New York: International Peacebuilding Academy.

UNDP (1994) 'Human Development Report 1994', Oxford: Oxford University Press.

United Nations (2004) 'Humanitarian Appeal 2004', www.un.org/depts/ocha/cap/appeals.html. Accessed: 15.02.2019.

United Nations (2010) 'United Nations Peace Building: An Orientation', New York: United Nations Peace Building Support Office.

United Nations Environment Programme (2009) 'From Conflict to Peacebuilding: The Role of Natural Resources and the Environment', Nairobi: United Nations Environment Programme.

United Nations General Assemble (2004) 'United Nations Secretary-General's High-Level Panel on Threats, Challenges and Change', New York: United Nations General Assembly.

United Nations Security Council (2007) 'Statement 2007/2 by the President of the Security Council', New York: United Nations Security Council.

Yilmaz, M.E. (2007) 'Intra-State Conflicts in the Post-Cold War Era', International Journal on World Peace, 24(4): 11-33.

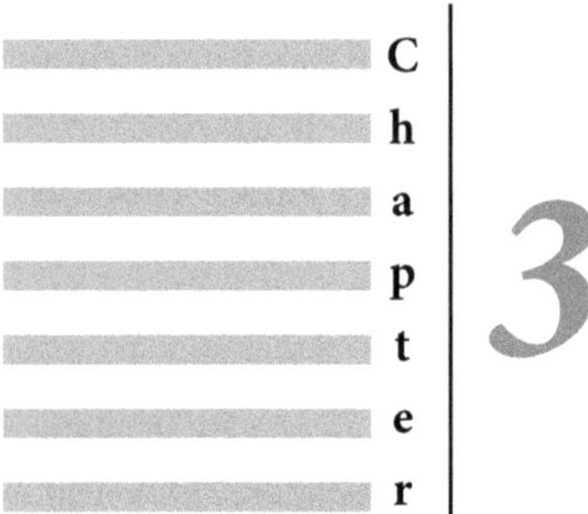

Youth Participation in Armed Conflict – A Theoretical Analysis

In today's world there are about 740,000 people who die and are affected as a result of armed violence each year. Armed violence erodes governance and peace whilst slowing down achievement of the Millennium Development Goals (MDGs). Armed violence can have a significant effect on security and development just as it has in societies that are affected by war or civil war. An armed violence agenda therefore includes a wide range of countries, cities and citizens whose development and security are under threat. It refers to the use or threatened use of weapons to inflict injury, death or psychological harm. It is critical that donors focus on youth because they are the largest and potentially most significant population in the developing world where approximately 1.3 billion youth live in developing countries (World Bank, 2007). A majority of them lack basic education, marketable skills, decent employment and opportunities for positive engagement in their communities. While most youth do not engage in significant or repeated acts of violence, evidence suggests that out of school and unemployed or under employed youth are at greater risk of becoming perpetrators – and victims – of violence and crime, along with youth who suffer from economic and social deprivation marginalization, neglect and abuse (Social Development Direct, 2009).

Youth involvement in conflict and violence is a global problem, yet research and interventions tend to be isolated to specific manifestations like participation of children in war, inter-group tensions in the aftermath of war, street living, inter-group fighting and inter-personal aggression or exclusion. Previous analyses have focused on specific manifestations of conflict, in part because conflicts are grounded in different geopolitical and economic situations that are interesting to researchers in different academic disciplines (Daiute, Beykont, Higson-Smith and Nucci, 2006). Research on young people's involvement in armed conflict has for example been studied primarily by anthropologists and therapists in Africa, the Middle East, South America, and the Balkans (Apfel and Simon, 1996; Bret and McCallin, 1998; Machal, 1996),while psychologists in the United States and Europe have done extensive research on physical and psychological conflict among peers (Elliot, Hamburg and Williams, 1998) and a range of social scientists have studied children living in the streets in nations with struggling economies (Brown, Larson and Saraswathi, 2002).

In today's world, children and youth grow up in circumstances where armed violence has become a norm within families, communities or states. Emerging trends in contemporary conflict that has an effect on youth include the increasing proximity of violence to the lives of young people and the eroding of boundaries between different kinds of violence (UNICEF, 2007). In the developing world young people have easy access to firearms that are available at cheap rates, poorly regulated and often sold illegally which increases the possibility of armed violence and also hinders peace building and humanitarian assistance. These weapons are easy for youth to learn to use and carry. For example, more than ninety percent of young people involved in conflict in a variety of roles in Liberia, Guinea and Sierra Leone had access to weapons (Florquin and Berman, 2005). Where there is easy access to weapons and the increased familiarity of youth with such small arms, has the capacity to sustain a culture of inter-personal and gang related violence not only in societies that have been stable or at peace but also in fragile and post-conflict societies (OECD, 2011). Today's actors in armed conflicts are often non-state actors who use non-traditional forms of warfare where such wars are intra-state rather than inter-state. Frequently, they adopt strategies that bring the battle more immediately to the civilian populations and into the lives of millions of youth. A state's use of paramilitary and proxy forces increases youth vulnerability because these forces are less accountable to the government or the public. Youth are increasingly used as perpetrators or accomplices in terrorist acts and in some places have increasingly come under suspicion and suffer severe abuses when detained. Groups that pursue violent methods of achieving their goals are often motivated with criminal intentions, ideological and political motivations which makes it difficult to draw a clear picture of the groups' intentions. For example, armed conflict that began over political grievances can be furthered by opportunistic greed. Mischaracterizing violent groups can exclude them from conflict resolution dialogue and demobilization (UNICEF, 2007). The result of the realities described above is that young people are often involved in armed violence simultaneously as perpetrators, victims and witnesses. Young boys and girls are frequently victims of violence where boys and young men are at the risk of death and homicide due to conflict while girls and young women face increased risk of sexual violence especially in situations of armed conflict. Homicide on youth and survivors of violence often contribute to global concerns of premature death, injury that in many cases could lead to disability. It could also have a serious impact sometimes lifelong in terms of their behavior and social functioning which in turn has an effect on the victims' families, friends and communities. Young peoples' involvement in violence greatly increases the costs related to health and welfare services, limits productivity, diminishes value for property and disrupts essential services that weakens the fabric of society (WHO, 2004).

One of the challenges that is recurrent in youth work is defining who they are. In this context, for example, people in the west and among international agencies, there is a strong tendency to use age range to define the category of youth. In this common or more general approach there are atleast four problems. First, the age ranges continue to differ where a common range is 15-24 which is suggested by UNICEF and others. However, there is slight variation amongst international agencies such as Save the Children which specifies an age range of 13 to 25 (Sommers, 2001). Secondly, definitions of youth often overlap with the much more common age based definitions of child and adult. One of the common and widely accepted notion in the west and the United Nations is a person's 18th birthday which marks the separation point of moving directly from the status of child or minor (ages 0-17) directly into adulthood (18 and above). Thirdly, concepts of youth, adolescence and even what constitute being young also varies. Adolescence is generally assumed to be a subset of the category of youth. An official of the World Health Organization (WHO) described the complications that come with such overlapping assumptions - "adolescents are 10-19 years old and youth are 15-24 and young people are 10-24 years old" (Lowicki and Pillsbury, 2000). Finally, in many parts of the world definitions of youth are confused by the fact that the idea of youth may not be determined by age. The period of youth is often considered as the time of passage between childhood and adulthood or as the period between puberty and parenthood which is a biological marker. Male and female initiation rites mark the passage between childhood and adulthood in some cultures while in others females are considered as youth only before marriage which in some cultures occur in an early age and have children soon after reaching puberty which results in they becoming young adults at a very early stage of life and no longer youth or even children (Sommers, 2001). In Darfur, the idea or the concept of youth as a stage of development is unknown. Females are considered girls until they reach puberty at which point they are known as women. The social status of women changes when motherhood is attained and its effects are far greater vis-à-vis fatherhood (USAID, 2006).

Policy makers and practitioners have tended to conduct separate analysis for different forms of collective violence – political, criminal and ideological – in which youth are involved. A close examination of the evidence suggests that the underlying psychological factors that influence voluntary youth participation in different types of violent groups are similar, despite the different contexts in which youth participate. The developmental tasks of adolescence include solidifying a set of values that guide behavior, responding more to their peers than adults, achieving independence from adults both emotionally and financially and becoming members of the community (Erikson, 1968). In conflict environments coming of age becomes particularly difficult for young people to complete the period of transition as community structures and state functions such as health care and education are disrupted. Young people who belong to marginalized groups often find it more attractive to join violent groups as it seemingly offers a faster transition into adulthood as well as protection and opportunities for advancement economically along with adventure. It also offers young people a sense of identity and affinity, develop respect and status among their peers and the community and opportunity to take revenge on other groups. Proximity, opportunity and familiarity are the reasons why young people join a particular type of group and it is not based on a fundamental reflection over differences in the psychological motivations of young people. Whilst men predominantly involve in armed violence, women play support roles by engaging in perpetrating acts of violence themselves (OECD, 2011).

There are fundamental contradictions that are present in the literature on youth. The depiction of youth is generally that of passive victims of trauma or as agents of active security threats. These separate depictions have separate origins. Graca Machel's (1996) landmark submission to the UN General Assembly (UNGA) – the Report on the Impact of Armed Conflict on Children describes how war traumatizes children. The report describes the devastating effects of war on children. Eversince, Machel has conducted several studies that have reemphasized the fact that although some children may have been spared direct experience of violence in armed conflicts, yet they still suffer deep emotional distress. She concluded that all children who have lived through conflict situations would require psychological support (Machel, 2001).

These popular depiction of young men as threats to security for the most part has been a western notion. Robert D. Kaplan characterized male youth living in West Africa as "out of school, unemployed, loose molecules in an unstable social fluid that threaten to ignite" (Kaplan, 1996). These descriptions as menacing as they are, have also been supported by Samuel P. Huntington who has argued that societies are particularly vulnerable to war when people belonging to the age group of 15-25 (youth) comprise atleast 20 percent of the total populations. This thesis showcases the demographic dangers that could be created through 'youth bulges' which is defined as extraordinarily large youth cohorts relative to the adult populace (Urdal, 2004).

According to this view of young people, male youth may not be the only ones to fear. Hendrixson highlights the counterpart image of the aggressively heterosexual angry young man, whose presence accentuates the implied violence and is the passive, veiled young woman, whose presence accentuates the implied violence and menace of all youth (Hendrixson, 2004). The author argues that the passive young woman image lumps together southern and Muslim women into a single figure hidden behind a veil interpreted as a sign of women's unfreedom *(ibid)*. As depicted by adherents of the youth bulge threat, female youth are menacing as they are, effectively, the principal source of the bulge. Hendrixson argues viewing female youth mainly as potential mothers which reinforces the notion that young southern women's fertility is responsible for population growth and more specifically for the rise in the number of young male terrorists *(ibid)*.

The literature that touches on youth and violent conflict focuses on analyzing the reasons why young people join the fighting. It is often remarked that war would not be possible without youth as combatants of any war in any part of the world. Why is this the case? Are young people by the mere fact that they are young and energetic possess a tendency towards violence? Therefore, does a large proportion of young people in society per se mean a warning sign for trouble? Do young people fight for their own cause, or are they mobilized into war by others or do young people fight to change the existing conditions of their particular grievances and if so what are these grievances?

Most of the analyses on youth and violent conflict are produced by working backwards, that is, by analyzing the motivations of young people that are or have been fighting and generalizing these motivations as if they were applicable to the whole youth cohort in a particular context. Here an important question arises, what about the majority of youth who do not participate in violent activities? As Nicholas Argenti (1998) puts it, with specific reference to Africa:

> *'The remarkable thing is not why some of Africa's youth have embraced violence, but why so few of them have"*

It is also interesting to note that most of the literature is gender-biased. Although the necessity of incorporating a gender dimension is generally acknowledged, theories on youth and violence still implicitly or explicitly refer to young males. Literature on the victimhood of youth often focuses on young women, but it is the young males that are viewed as threats to security and thus emblematic of the societal problems resulting from the youth crisis. Girls tend to disappear from the mainstream literature on youth and violent conflict, except when in the role of victims (Bureau for Crisis Prevention and Recovery, 2005).

The focus on youth as an area of concern for the situation of youth emerged with strength in the UN developmental discussion between the years 1996 and 2001, according to the UNDP Youth and Conflict Report. Interest was initially due to the growing recognition of the challenges that young people face when returning from armed conflict and attempting to integrate into communities and find jobs or access to education. The UN view was that young people are disproportionately affected by violent conflict, both as victims and as active participants.

The focus on youth as an area of concern for the situation of youth emerged with strength in the UN developmental discussion between the years 1996 and 2001, according to the UNDP Youth and Conflict Report. Interest was initially due to the growing recognition of the challenges that young people face when returning from armed conflict and attempting to integrate into communities and find jobs or access to education. The UN view was that young people are disproportionately affected by violent conflict, both as victims and as active participants. However, youth have been primarily viewed as potential perpetrators, due to the influence of two intertwining agendas. The most influential idea driving an increase in youth in the last decade has been the youth bulge. The idea has been most closely associated with the research conducted by Henrik Urdal for the World Bank (Maytok, Senehi and Byrne, 2011). Caught between childhood and full adulthood, youth are often even more undeserved than children. Young people while struggling with their own identity often are in a situation that makes them witness the collapse of the social fabric. It is held from a conflict perspective that idleness, and especially the realization of a lack of future prospects In terms of employment and lack of educational opportunities represent not only social problems, but may further turn youth into those that are prone to be recruited into rebel armies and violent movements (World Bank, 2011).

Unemployed, uprooted, alienated young people and more particularly young men who have very limited opportunities for proper employment and positive engagement in society are a ready pool of recruits for armed groups or groups that have violent agendas. Several categories of young people appear to be particularly at risk - unemployed university graduates, young people who have moved from rural to urban areas and youth who have lived through internal conflicts. Young people often participate in violence because of limited avenues for constructive political participation. Equally, youth prioritize employment and as they put it 'if you give us something to do, we won't have any need to disrupt life'. In many of these societies, youth are marginalized and have little voice in the political processes, which in large part contributes to their frustration. Too often government and elected officials do not engage young citizens or encourage them to participate in politics. When young people do try to express their political wishes, government officials often overlook concerns that are specific to youth such as education and employment. The isolation of youth and the perception that their problems go largely unnoticed leads to frustration and tragically can lead to violence and conflict.

Despite the limited amount of systematic evidence on the reasons for youth engagement in violence, there are a number of overarching theories in the literature that examines their participation in violence. These are variously based on economic, biological, social and political analysis. One of the frequent suggestion is that exceptionally large youth cohorts or the so called youth bulges, often puts countries in situation that is more susceptible to political violence. The term youth bulge is a misnomer, although few authors use the same definition of youth bulge, almost all researchers measure it as the number of young people who are generally in the age grouping of 15 and 24. A bulge is literally defined as an irregular swelling (Abate, 1998), should be visible in the young adult section of the age pyramid. In the study of civil wars there are two prominent theoretical frameworks which argue that youth bulges potentially increase both opportunities and motivations for political violence. Using a time-series cross-national statistical model for internal armed conflicts for the period of 1950 and 2000 and in the case of event data for terrorism for the years between 1984 and 1995, these claims have been tested empirically. Jack. A Gladstone argues that youth have played a prominent role in the English Revolution to the revolutions of 1848 which suggests that youth bulges have been historically associated with times of political crisis. (Gladstone, 1991, 2001). A historian has even linked economic depression hitting the largest German youth cohorts to the rise of Nazism in Germany in the 1930s (Moller, 1968) Generally it has been observed that young males are the main protagonists of criminal violence (Neapolitan, 1997; Neumayer, 2003) as well as political violence (Mesquida and Weiner 1996; Elbadwi and Sambanis, 2000). It has been suggested that young men are more aggressive due to high male sex harmone levels (Goldstein, 2001; Hudson and den Boer, 2004).

Claims that youth bulges can be one of the causes for political violence has a long history (Moller, 1968; Choucri, 1974) and the issue has received increasing attention over the past decade which has following the more general debate placed its emphasis over security implications that is a result arising from population pressures and scarcity of resource. In his work 'The Coming Anarchy', Robert Kaplan argues that anarchy and the crumble away of nation states in the future can be attributed to factors that are related to demography and the environment (Kaplan, 1994). More recently, youth bulges have become a popular expansion for current political instability in the Arab World and for recruitment into international terrorist networks. In a background article surveying the root causes of the September 11, 2001, terrorist attacks on the twin towers in the United States, Newsweek editor Fareed Zakaria argues that youth bulges combined with slow economic and social change has provided the foundation for an Islamic resurgence in the Arab World (Zakaria, 2001). Although there is strong interest that is evinced among policy makers and academicians, few rigorous empirical studies have been conducted to establish the relationship between youth bulges and political violence. Two prominent quantitative studies on causes of civil war onset, Fearon and Laitin (2003) Collier and Hoefler (2004), initially included measures of youth bulges as one among a high number of variables. A third rigorous study including youth bulges among an even higher number of regressors, the State Failure Task Force Report (Etsy et.al., 1998), finds some effect of youth bulges on ethnic conflict.

The focus on the literature on youth bulges has been in particular on spontaneous and low intensity unrest such as non-violent protest, riots and rebellion. Henrik Urdal (2006) argues that the concept of youth bulges are indeed relevant for understanding and explaining the occurrence of more organized forms of political violence such as internal armed conflicts or ethnic conflicts. Following the September 11 attacks, there has been an increase in the interest and attention on youth bulges as a cause for political violence or a possible explanation for terrorism and increased global insecurity can be noticed. The World Bank in its report on 'the economics of conflict', found that large proportions of young men in a country may increase the possibility of conflict (World Bank, 2011). Strain on social institutions such as the labour market, the educational system thereby resulting in violent conflict are a result of youth bulges has a long history (Choucri, 1974; Moller, 1968). However, the issue has received wide attention over the past decade following the debates on security implications on pressures of population and resource scarcity. Despite its long history, it has been after the end of the Cold War that the claim youth bulges and other demographic factors have become more important in the context of conflicts. In his statement to the Senate Committee on Intelligence held in the year 1997, the Director of Defense Intelligence Agency, Patrick M. Hughes pointed out that although there is no peer competitor to the US after the end of the Cold War, the world continues to remain a very dangerous and complex place (Hughes, 1997). In his list of conditions that are possible factors for making the world a very dangerous place, the existence of youth bulges was placed at the top. Thus, the idea of a large proportion of young people within a society that may be a cause for violent conflicts is not new. In this context, many scholars, researchers and policy makers have already examined the question of how and to what extent the two phenomena can be related or linked (example see Choucri 1974). However, scientific explanations over the link between the two began in the 1990s. The term youth bulge was coined by Fuller (1995). The phrase youth bulge is generally used to describe a situation in which those in the age group of

15-24 exceed 20 percent of the total population and the share of people belonging to the age group of 0-14 - often referred to as children bulge which serves a predictor for future youth bulge – is higher than 30 percent. The theory of youth bulge thus predicts that societies can be characterized as having a youth bulge while simultaneously facing lack of resources and in particular opportunities for young people to hold prestigious positions in the society. While the idea of youth bulge is not entirely new there were similar theories that were suggested earlier by Moller (1968) and Bouthal (1970). The application of the theory of youth bulge has been in particular, common with regard to the Muslim world, which is in recent times facing an unprecedentedly high levels of youth bulge. Additionally, Samuel P. Huntington endorsed the idea of Fuller's theory of youth bulge in his influential book, 'The Clash of Civilizations and the Remaking of the World Order'. Until the publication Heinsohn's (2003) publication of 'Sons and World Power: Terror in the Rise and Fall of Nations', the idea of youth bulge was rather ignored in continental Europe and in Germany. Heinsohn adopts Goldstone's (1991) approach and argues that it is the excess of young male adults within any given youth bulge that helps explain most periods of internal and exported social unrest in human history such as civil wars, terrorism, imperialism, genocide and ethnic conflicts. Heinsohn in similar fashion defines youth bulge as when the ratio of adequate positions demanded by succeeding sons is substantially smaller. He further adds that these surplus sons are vulnerable to all forms of just violence that could be a result of indoctrination by political and various sorts of religious extremists. Heinsohn is also careful to point out that any mono-causal explanation of any sort of major social unrest is a result of youth bulges may not be true in all cases. Infact, he argues, history's most outrageous atrocities – *e.g.* Stalin's launch of a command and control economy in the Soviet Union which resulted in the death of about 30 to 60 million people, and the holocaust in Hitler's Germany after the start of the Second World War in 1939 – cannot be explained as a result of youth bulges (though Russia displayed a massive youth bulge from 1897 to the years immediately before these young Bolsheviki staged the 1917 October Revolution; Germany's last youth bulge – with a 35 percent population share – was recorded between 1919 and 1933, and gave a big boost to the national socialist movement's drive to power in 1933, however, after 1939 most German soldiers were from families with less than two sons). Heinsohn however regarded that youth bulges were atleast an important contributing factor in most outbreaks of social unrest – regardless of the type of conflict. In order to prove his hypothesis, he provides a list of the world's largest countries (in 2003) ranked according to the absolute number of children under the age of 15. In his list the top ten countries that were listed in rank order are India, China, Indonesia, Pakistan, Bangladesh, Mexico, Brazil, USA, Nigeria and Ethiopia. Seven of the top ten countries had over thirty percent of 0-15 year olds with Ethiopia, Nigeria and Pakistan having the largest ratio (47 percent, 44 percent and 40 percent). In contrast, Brazil, China and the USA are not characterized as having a youth bulge (28 percent, 24 percent and 21 percent). Finally, he shows that of the 67 countries in his list that were affected by a youth bulge at some period after 1945, 65 had experience major social unrest, with all major religions (except Buddhism) and political ideologies having been involved (Schomaker: 2013).

Some theorists have proposed that youth cohorts may develop a generational consciousness and more particularly due to the awareness of belonging to a generation of extraordinary size and strength, enables them to work or act collectively (Braungart, 1984; Feuer, 1969; Goldstone, 1999). However, they also suggest that violent conflict between groups that are divided by age are rare. This generational approach has serious shortcomings in terms of the explanatory power of the relationship between youth bulges and violence. The development of generational consciousness may help in the formation of youth movements that in most cases function as identity groups. Identity groups are an important necessity for collective violent action to take place. However, it is not necessary that identity groups are generation-based so that youth bulges would result in or increase the likelihood of armed conflict. Furthermore, the generational approach falls short of providing convincing explanations for the motivations for youth rebellion nor does it provide a substantial explanation for the opportunities of conflict. It is clear that if large youth bulges who have a common generational consciousness, would produce conflict, we would have witnessed a lot more violent youth revolts (Urdal, 2004).

> *'Countries that experience youth bulges are more likely to experience domestic armed conflict than countries that do not'*

Youth bulges increases unemployment because there is the availability of labor substantially when entering the labor market. Unemployment is believed to cause grievances, and especially when expectations are raised through expansions in education. Similarly, grievances arise if there are possibilities to influence the political system and attain elite positions are limited. The idea that an exceptionally large share of young people or a large youth cohort size will lead to social unrest can be based on several transmission channels. One of the most prominent approaches, focusing on the supply side is based on the assumption that this phenomenon increases the supply of cheap labor (Urdal, 2006). Limited employment opportunities – caused by limited absorption capacity of the labor market in the face of a sudden surplus of labor – or the decreasing relative (male) wages can result if more potential workers compete for only few jobs (Easterlin, 1987; Niang, 2010). Dissatisfaction and frustration (and related consequences) among young people are likely to be the consequence, in particular if compared to their parents' smaller generation – potential lifetime income is perceived to be limited. In addition, these arguments are backed by several studies in economic demography which suggest that the alternative costs of individuals who are part of an existing youth bulge are lower in terms of economic and social fortunes (Urdal, 2006), compared to members of smaller cohorts.

Given the importance of youth employment, the links between youth demography and youth unemployment are worthy of analysis while looking at the problem in terms of a policy issue. The International Labor Organization in the year 2013 estimated a global youth unemployment rate of 12.6 percent with an estimated 73 million young people unemployed (ILO, 2013). Youth unemployment tends to be substantially higher vis-à-vis adult unemployment rates in all countries the ratio of youth unemployment to overall adult unemployment was estimated at 2.7 percent that is similar to the ration in recent years (ILO, 2013). Many discussions of youth employment talk about the demography of youth populations. Rapid population growth in many developing countries in the 1960s and 1970s have produced large youth populations (Lee, 2003; Lam and Marteleto, 2008; Lam 2011; Assad and Levison, 2013). It is thus important to consider the potential impact of large and growing youth populations on youth employment and other labor markets.

Youth bulge has frequently been mentioned in discussion of the Arab Spring (LaGraffe, 2012). One of the mechanisms frequently mentioned for a link between the youth bulge and political unrest is that large youth cohorts may contribute to high youth unemployment. Direct evidence on a link between the relative size of the youth population and youth unemployment is quite limited, however, especially in developing countries. Studies on the relationship between cohort size and labor market outcomes in high income countries have often found that larger cohorts experience worse labor market outcomes. A large literature focused on the early labor market experience of the large baby boom cohorts that entered the labor market in the 1960s and 1970s in North America and Europe (eg. Welch, 1979; Berger, 1985, loom et.al. 1987; Zimmerman, 1991). The broad consensus of these studies was that larger cohort size was associated with some combination of lower entry level wages and higher unemployment relative to older workers, with differences across countries in the extent to which wages or unemployment showed the largest effects of cohort size.

Korenman and Neumark (2000) used data for 15 OECD countries from 1970-94 to combine variations across countries with variation across time to examine the impact of cohort crowding on youth labor markets. Their estimates suggest that a higher youth share of the working-age population leads to a higher youth unemployment rate relative to adult unemployment. Using state-level data for the United States, they found the surprising result that an increase in the youth share of the working-age population reduces both the youth unemployment rate and the prime-age adult unemployment rate. Drawing on predictions from a search model of the labor market, they attributed this result to the fact that high fractions of young people in the labor force lead to increased labor market flexibility. There has been relatively little research analyzing the impact of cohort size on labor market outcomes in developing countries. Behrman and Birdsall (1988) found that being in a large cohort had negative impact on labor market outcomes of unskilled men in Brazil. Lam (2006) and Assad and Levison (2013) showed that the youth proportion of the working-age population has declined in many developing countries, the result of rapid fertility declines. Fares et.al. (2006) analyzed data for 93 countries and found little evidence that larger youth cohorts had worse labor market outcomes.

The possibility of a young person becoming a victim or perpetrator of violence can be attributed to risk factors for youth armed violence. Preventing violence involves direct efforts to remove or reduce risk factors, as well as harnessing the indirect effects of other policies and programmes that may reduce exposure to underlying causes and risk factors. No single factor explains why a person or group is at a high or low risk of such violence. Instead, violence is an outcome of interaction among risk factors at four levels (i) individual (including biological factors), (ii) relationship, (iii) community and (iv) societal (WHO, 2002). The following risk factors for youth violence has been documented in a variety of socially and culturally-distinct settings. Individual factors include traits such as hyperactivity, impulsiveness, poor behavioral control and attention problems; a history of early aggressive behavior; early involvement with drugs, alcohol and tobacco; anti-social beliefs and attitudes; low intelligence; low commitment to school and failure; and exposure to violence and conflict in the family. Additionally, studies show that drunkenness is an important and immediate situational factor that can precipitate violence. Relationship factors refer to family, friend, intimate partner and peer relationships, and include poor parental supervision of children; harsh or inconsistent disciplinary practices; witnessing violence or experiencing abuse during childhood; low levels of attachment between parents and children; low parental involvement in children's activities; parental alcohol/substance abuse or criminality; experiencing parental separation or divorce at a young age; poor family functioning; coming from a single-parent household; and low socio-economic status of the household. Associating with delinquent peers is also an important factor for youth violence. Community factors include high residential mobility, high unemployment, high population density, social isolation, proximity to drug trade, ease of access to alcohol, and weak social welfare policies and programmes in schools. Societal factors include rapid social change, economic and gender inequalities, social policies that create and sustain or increase economic and social inequalities, poverty, weak criminal justice systems that allow the excessive use of force by police with impunity, the availability of firearms and social and cultural norms that support violence. Risk factors can also be considered in terms of (i) factors that push youth to engage in violence, (ii) factors that pull or attract youth towards violent acts and (iii) factors that may trigger violence (Social Development Direct, 2009). There is a strong interplay and causal relationship among the three factors.

The literature on youth armed violence outlines a number of different factors that may push youth into violence or towards joining violent or extremist groups. These include poor living conditions, devastation of family and social structures from disease, limited or unequal access to education and vocational training, lack of employment or livelihood opportunities, social exclusion and inequality, weak political participation and decision making power, previous exposure to violence and lack of public safety and security. Recent research suggests that political, social and economic exclusion can prevent youth from completing the key steps their societies require to achieve adulthood, effectively blocking the transition of many young people to adulthood and resulting in frustration and discontent which may be expressed in acts of violence and leave them more vulnerable to coercion. Young people fight because they are forced to – either by physical abduction, or because of a lack of other alternatives for survival. The corollary of this is that young people are not really responsible for their choice to fight and should be treated as victims rather than as perpetrators. This perspective is found especially in the burgeoning literature on child soldiers, produced largely by aid agencies and NGOs and based, to a large extent, on witness accounts of former child soldiers. These reports are relevant for this chapter, because, although focusing on underage combatants, they also consider young people that are over 15 years of age (usually in the age range of 15-18). Research also suggests that there are a number of factors that pull individuals towards armed violence, or networks or groups that advocate violence. These factors include a need to uphold values and ideology, a sense of inclusion or identity, a sense of safety or protection, a means to obtain material and non-material benefits, a sense of adventure and excitement, a sense of structure to life and the opportunity for a voice in their societies. Certain situations may trigger violent behavior. On an individual level, violence by a young person may be sparked by events in their lives, such as mistreatment or detention by the police, loss of a job or failure to find one, rejection by a peer, partner or family member, substance abuse, or emotional trauma. On a group level, the decision of a group to engage in violence can be sparked by public events such as acts that citizens view as coerced or accomplished by fraud and deceit of public officials, public denigration of an ethnic or religious group, abuse by security forces, policy changes or economic crises (OECD, 2011).

Criticisms on the Youth Bulge

Despite widespread support, the principles of contemporary youth bulge theory have not gone unchallenged. The theory and its arguments provide a hotbed for criticisms and debate. For example, Paul Collier and Anke Hoeffler (2001) failed to find any significant relationship between large youth cohorts and armed uprisings. Similarly, the US sponsored State Failure Task force has also failed to find significant effects of youth bulges as a cause for violent conflict (Etsy *et al.*, 1998). Further, criticisms on the youth bulge theory stems from doubt that demographic factors in and of themselves are strong enough to indicate a definitive outcome of violence (Holmes, 2001). And although youth bulge theory qualifications have suggested that certain measures of economic development or regime types can deepen youth bulge risks, they have not offered much insight into the relationships between these complex factors (Barker and Ricardo, 2006). While poverty, inequality and political disenfranchisement persist in the lives of many young people who do engage in violent acts, these indicators have been shown to have weak explanatory power for causing violence (Goldstone, 2001). Furthermore, for all youth who do engage in violence amidst these conditions, there are many more who do not (Barker and Ricardo, 2006). Hence, there is a dynamic interaction missing from the youth bulge argument that warrants further examination. Researchers continue to discount violent behavior in young men as natural or inherent. As researchers Barker and Ricardo (2006) note – the majority of violent behavior is explained by social factors during adolescence and childhood. 'Boys are not born violent – they learn to be violent'. This point is strongly reiterated by studies that have found that adolescent males with high T-levels of testosterone are not necessarily more violent than others, but rather that they are more easily influenced by peer groups. In this line of reasoning, if adolescent males are surrounded by peers engaged in delinquency, they are likely to copy that behavior. However, if they are around those engaged in positive activities, they are likely to become leaders (Sommers, 2006).

These conclusions introduce an argument that testosterone may be infact related to leadership rather than to anti-social behavior (Rowe et.al., 2004). Arguments such as these compel many critical theorists seeking to understand how youth bulges can become violent. Youth bulge theory mistakenly discounts that young men and women are thinking, feeling and rational agents. Young people ultimately choose whether or not and under which circumstances to engage in violent behavior, and those choices are often shaped through formative, repeated socio-cultural processes (Butler, 2009). Furthermore, the differences in socialization between young men and women in most cultures suggest that these processes may influence the disproportionate participation of young males in violent acts. Specifically, many researchers are interested in the ways in which the theory reflects racial, gender and age discrimination (Hendrixson, 2004). Some arguments claim that youth bulge theory helps give momentum to a type of orientalism (see Said, 1978) in its positioning of young men of color as security threats. As Anne Hendrixson has pointed out, youth bulge theory frames these young men as angry terrorists, while their female counterparts are positioned as little more than veiled victims (Hendrixson, 2004). She goes on to argue that the effects of these frames and representations are used to justify hardline western military interventions and repressive population control methods in youth bulge countries (ibid). Others have added that the direction of foreign assistance toward repressive mechanisms and away from the creation of socio-economic opportunities for young people has been an exceedingly detrimental impact of the theory of youth bulge theory (Sommers, 2006).

The suggestion follows that youth bulge theory, at the same time as it frames young people in LDCs as threats, positions young people in countries of higher economic development as perhaps less violent, or even as scarce assets (Hendrixson, 2004). Indeed, one finds a stark difference in the framing of young people between north and south. Throughout the developed world, the concept of demographic gift has been used to describe the economic boom seen in populations that have especially large youth bulges of working-age citizens relative to dependent populations (ages 0-4 and 64+). In their study of the economic effect of the demographic gift in East Asia, David E. Bloom and Jeffrey G. Williamson (1998) found that population dynamics may have accounted for almost half of the region's economic miracle. Similarly, Paul Kennedy (2003), has argued that in order for Europe to achieve greater influential power on the international playing field, it must invest in producing more young people. The demographic gift and the youth bulge are two distinct demographic scenarios, the former requires a past demographic transition from high to low fertility rates, does not take place automatically and requires sufficient policies to harness real economic benefits (Bloom and Willaimson, 1998; Jackson and Flemingham, 2002). Nevertheless, the beneficial, constructive and profit bearing potential of youth cohorts illuminated through the demographic gift theory stand in stark contrast next to youth cohorts as they are framed in youth bulge societies. Moreover, Tarik Yousef and Navtej Dhillon (2007) have conjectured that the MENA regions youth bulge could be transformed into a demographic gift if, over the next 20 years, youth participation in the labor forces were to rise from 39 percent to 67 percent (Similar to that of youth labor force participation in East Asia in 2005) and youth unemployment were reduced by half. The authors warn that this window of opportunity to harness the MENA's demographic potential is anticipated to close by 2045, and argue that taking advantage of the demographic dividend will require enhanced understanding of the aspirations and expectations of young people. Finally, a critical weakness of youth bulge lies in is lack of focus on non-violent youth action, which notably reflects the choices of most of the youth in youth bulge societies (Barker and Ricardo, 2005). Mark Boren has documented numerous examples of how masses of young people have historically not only mobilized themselves to disrupt various social and political systems, but how they have also acted as meaningful and positive agents of change, often through non-violent means (Boren, 2001).

REFERENCES

Apfel, R.J. and Simon, B. (1996) '*Minefields in their hearts: The Mental Health of Children in War*', New Haven: Yale University Press.

Argenti, N. (1998) 'Air Youth: Performance, Violence and the State in Cameroon', *Journal of the Royal Anthropological Institute*, 4(4): 753-781.

Assad, R. and Deborah, L. (2013) '*Employment for Youth – A growing Challenge for the Global Community*', Background Research Paper of the United Nations High Level Panel on the Post 2015 Development Agenda', Minnesota: University of Minnesota Press.

Baker, G. and Ricardo, L. (2006) 'Young Men and the Construction of Masculinity in Sub-Saharan Africa: Implications for HIV/AIDS, Conflict and Violence', in Bannon, I and Marion, L.L. (eds.) *The Other half of Gender: Men's Issues in Development*, Washington D.C: World Bank.

Behrman, J.R. and Birdsall, N. (1988) 'The Reward for Good Timing: Cohort Effects and Earnings Function for Brazilian Males', *Review of Economics and Statistics*, 70(1): 129-135.

Berger, M.C. (1985) 'The Effect of Cohort Size on Earning Growth: A Re-examination of the Evidence', *Journal of Political Economy*', 93(1): 561-573.

Bloom, D.E., Freeman, R.B. and Korenman, S. (1987) 'The Labor Market Consequences of Generational Crowding', *European Journal of Population*, 3(1): 131-176.

Bloom, D.E. and Williamson, J.G. (1998) 'Demographic Transitions and Economic Miracles in Emerging East Asia', *World Bank Economic Review*, 12(3): 419-455.

Boren, M.E. (2001) '*Student Resistance: A History of the Unruly Subject*', New York: Routledge.

Bouthal, G. (1970) '*L'infanticide Differe*', Paris: Hatchette.

Braungart, R.G. (1984) 'Historical and Generational Patterns of Youth Movements: A Global Perspective', *Comparative Social Research*, 7(1): 3-62.

Brett, R. and McCallin, M. (1998) '*Children: The Invincible Soldiers*', New York: United Nations Children's Fund.

Brown, B.B., Larson, R.W. and Saraswathi, T.S. (2002) '*The World's Youth: Adolescence in Eight Regions of the Globe*', New York: Cambridge University Press.

Bureau for Crisis Prevention and Recovery (2005) '*Youth and Violent Conflict: Society and Development in Crisis?*', Geneva: Bureau for Crisis Prevention and Recovery.

Butler, J. (2009) '*Frames of War: When is Life Grievable?*', London: Verso.

Choucri, N. (1974) '*Population Dynamics and International Violence: Propositions, Insights and Evidence*', Lexington: Lexington.

Collier, P. and Heoffler, A. (2004) 'Greed and Grievance in Civil War', *Oxford Economic Papers*, 56(1): 563-595.

Dhillon, G and Yousef, T. (2007) '*Inclusion: Meeting the 100 Million Challenge Flagship Report*. Middle East Youth Initiative: Wolfensohn Center for development at the Brookings Institution and Dubai School of Government.

Easterlin, R.A. (1987) 'Easterlin Hypothesis' in Eatwell, J., Millgate and Newman, P. (eds.) *A Dictionary of Economics*, Vol.2. New York: Stockton.

Elbadawi, I and Sambanis, N. (2000) 'Why are there so Many Civil Wars in Africa?: Understanding and Preventing Conflict', *Journal of African Economies*, 9(3): 244-269.

Elliot, D.S., Hamburg, B.A. and Williams, K.R. (1998) '*Violence in American Schools*', New York: Cambridge University Press.

Erickson, E.H. (1968) '*Identity, Youth and Crisis*', New York: Norton.

Etsy, D.C. *et al.* (1998) '*State Failure Task Force Report: Phase II Findings*', Mclean: Science Applications International.

Fares, J., Montenegro, C.E and Orazern, P.F. (2006) '*How are Youth Faring in the Labor Market? Evidence from Around the World*', World Bank Policy Research Paper, 4071.

Fearon, J.D. and Laitin, D.D. (2003) 'Ethnicity, Insurgency and Civil War', *American Political Science Review*, 97(1): 75-90.

Feure, L.S. (1969) '*The Conflict of Generations: The Character and Significance of Student Movements*', London: Heinemann.

Florquin, N. and Berman, E. (eds.) (2005) '*Armed and Aimless: Armed Groups, Guns and Human Security in the ECOWAS Region*', Geneva: Small Arms Survey.

Gladstone, J.A. (1991) '*Revolution and Rebellion in the Early Modern World*', Berkeley: University of California Press.

Gladstone J.A. (2001) 'Demography, Environment and Security', in Paul, F.D and Nils, P.G. (eds.) *Environmental Conflict*, Boulder: Westview.

Goldstein, J. (2001) '*War and Gender: How Gender Shapes the War System and Vice-Versa*', Cambridge: Cambridge University Press.

Heinsohn, G. (2003) '*Sons and World Power: Terror in the Rise and Fall of Nations*', Bern: Orell Fusli.

Hendrixson, A. (2004) '*Angry Young Men, Veiled Young Women: Constructing a New Population Threat*', Corner House Briefing No. 34.

Holmes, J.S. (2001) '*Radicalism: Is the Devil in the Demographics?*', New York Times, December 9, 2001.

Hudson, V.M. and Den Boer, A.M. (2004) '*Bare Branches: The Security Implications of Asia's Surplus Male Population*', Cambridge: MIT Press.

Hughes, P.M. (1997) '*Global Threats and Challenges to the United States and its Interests Abroad: Statement for the Senate Select Committee on Intelligence*', February 6, 1997, Washington D.C: Defence Intelligence Agency.

Huntington, S.P. (1996) '*The Clash of Civilizations and the Remaking of the World Order*', New York: Simon and Schuster.

International Labor Organization (2013) '*Global Employment Trends for Youth 2013: A Generation at Risk*', Geneva: International Labor Organization.

Jackson, N. and Flemingham (2003) '*The Demographic Gift in Australia*', Discussion Paper, School of Economics, Hobart: University of Tasmania.

Kaplan, R.D. (1994) 'The Coming Anarchy', *Atlantic Monthly*, 273(1): 44-76.

Kaplan, R.D. (1996) '*The Ends of the Earth: A Journey at the Dawn of the 21st Century*', New York: Random House.

Kennedy, P. (2003) '*Europe's Laggards Will Never Balance U.S. Power*', The Guardian, June 24, 2003.

Korenman, S. and Neuman, D. (2000) 'Cohort Crowding and Youth Labor Markets: A Cross-Sectional Analysis', in Blanchflower, D. and Freeman, R. (eds.) *Youth Employment and Joblessness in Advanced Countries*, Chicago: Chicago University Press.

La Graffe, D. (20102) 'The Youth Bulge in Egypt: An Intersection of Demographics, Security and the Arab Spring', *Journal of Strategic Security*, 5(2): 65-79.

Lam, D. and Leticia, M. (2008) 'Stages of the Demographic Transition from and Child's Perspective: Family Size, Cohort Size and Schooling', *Population and Development Review*, 34(2): 225-252.

Lam, D. (2011) 'How the World Survived the Population Bomb: Lessons from Fifty Years of Extraordinary Demographic History, *Demography*, 48(4): 1231-1262.

Lee, R. (2003) 'The Demographic Transition: Three Centuries of Fundamental Change', *Journal of Economic Perspectives*, 17(1): 167-190.

Lowickci, J. and Pillsbury, A. (2000) '*Untapped Potential: Adolescents Affected by Armed Conflict: A Review of Programs and Policies*', New York: Women's Commission for Refugee Women and Children.

Machel, G. (2001) '*The Impact of War on Children: A Review of the Progress Since 1996*', Geneva: United Nations.

Maytok, T., Senehi, J. and Byrne, S. (2011) '*Critical Issues in Peace and Conflict Studies: Theory, Practice and Pedagogy*', Maryland: Lexington Books.

Moller, S.H. (1968) 'Youth as a Force in the Modern World', *Comparative Studies in Society and History*', 10(1): 238-260.

Neopolitan, J.L. (1997) '*Cross-National Crime: A Research Review and Sourcebook*', Westport: Greenwood.

Neumayer, E. (2003) 'Good Policy Can Lower Violent Crime: Evidence from a Cross-National Panel of Homicide Rates 1980-97', *Journal of Peace Research* 40(1): 247-262.

Niang, S.R (2010) '*Terrorizing Ages: The Effects of Youth Densities and the Relative Youth Cohort Size on the Likelihood and the Pervasiveness of Terrorism*', Paper Presented at the Annual Meeting of the Midwest Political Association, April 2010, Chicago.

OECD (2011) '*Rethinking the Involvement of Youth in Armed Violence: Programming Note*', http://dx.doi.org/10.1787.9789264107205-en.

Rowe, R.C et.al., (2004) 'Testosterone, Anti-Social Behavior and Social Dominance in Boys: Pubertal Development and Bio-Social Interaction', *Biological Psychiatry*, 55(1):546-552.

Schomaker, R. (2013) 'Youth Bulges, Poor Institutional Quality and Missing Migration Opportunity – Triggers of the Potential Counter-Measures for Terrorism in MENA', *Topics in Middle Eastern and African Economies*, 15(1): 116-140.

Social Development Direct (2009) '*Youth Exclusion, Violence, Conflict and Fragile States*', London: Social Development Direct.

Sommers, M (2001) '*Youth Care and Protection of Children in Emergencies: A Field Guide*', New York: Save the Children.

UNICEF (2007) '*Children and Conflict in a Changing World*', New York: UNICEF.

Urdal, H. (2004) '*The Devil in the Demographics: The Effect of Youth Bulges on Domestic Armed Conflict, 1950-2000*', Social Development Papers: Conflict Prevention and Reconstruction Paper No. 14. Washington D.C: World Bank.

Urdal, H. (2006) 'A Clash of Generations? Youth Bulge and Political Violence', *International Studies Quarterly*, 50(1): 607-629.

USAID (2006) '*Youth and Conflict: A Brief Review of the Available Literature*', New York: USAID.

WELCH, F. (1979) 'Effects of Cohort Size on Earnings: The Baby Boom Babies, Financial Bust', *Journal of Political Economy*, 8(5): 65-97.

World Bank (2007) '*World Bank Report*', Washington D.C: The World Bank.

World Bank (2011) '*Children and Youth in Conflict*', New York: World Bank.

World Health Organization (2002) '*World Report on Health and Violence*', Geneva: World Health Organization.

World Health Organization (2004) '*Preventing Violence and Reducing its Impact: How Development Agencies Can Help* ', Geneva: World Health Organization.

Zakaria, F. (2001) '*The Roots of Rage*', Newsweek, 138(1): 14-33.

Zimmerman, K.F. (1991) 'Ageing and the Labor Market: Age structure, Cohort Size and Unemployment', *Journal of Population Economics*, 4(1): 177-200.

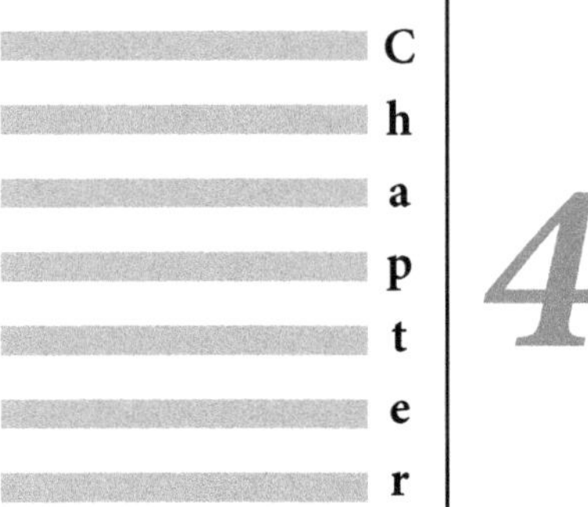

Youth in Peacebuilding

In areas that are affected by violent conflict programming for youth has seen a significant increase over the last decade. Organizations working with foreign assistance have recognized the need to engage this important demographic group and have increased direct assistance to young people or have begun efforts to mainstream attention to youth in their broader programs. This heightened sensitivity and expanded programming by international actors has increased the understanding that young people are active participants in rebuilding communities that have been torn apart by conflict. It has also highlighted opportunities for programmatic and policy-level challenges around effective projective design, coordination, funding, impact and accountability. These challenges encourage more sustained dialogue between agencies about the effects of conflict on youth and about how current practices fit within larger trends in youth programming in conflict affected areas, such as their assessment and evaluation.

A key lesson from modern world history is that young people act as engines of socio-political change, if not always, atleast its primary engineers. When effectively mobilized, youth provide the necessary energy and the mass power to get the wheels turning for divergent vehicles of change in their communities. Young people play major roles in conflict situations. Not only are they victimized by war, they are also manipulated and pulled in as combatants, ideologues, political thugs, *etc.* Opportunity for armed mobilization is immense, yet youth seeking non-violent roles are often left isolated and unsupported. Although there are efforts in recent times to engage in the peacebuilding process, there is clearly not much evidence that youth have been fully engaged in the process. Challenges and hindrances appear in many forms for the engagement of youth in peacebuilding. Are youth engaged in peacebuilding fully? Can we find a genuine commitment to engage youth in the peacebuilding process? This may seem an obvious question, but from a peacebuilding perspective it is crucial. The literature on youth engagement is littered with criticisms of superficial and tokenistic approaches, and in the field of peacebuilding, this is compounded by a tendency to characterize youth more as a risk to security than a resource for building peace and stability. Even where there is a genuine commitment to youth engagement, there are added complexities in situations where violent conflict exists. Youth cannot be understood as a homogeneous group that exists in isolation from conflict itself. Youth may be perpetrators as well as victims of violence, mobilized to fuel conflict as well as motivated to it. This presents significant challenges for genuine youth engagement in terms of identifying the multiplicity of youth perspectives on the conflict, the politics of who represents youth opinion and which youth organizations receive funding and resources. The greatest challenge in the process of peacebuilding is that it involves including politically opposed and marginalized groups into the dialogue process. While peace agreements might represent new working arrangements between political elites, relapses into conflict are common, especially where younger generations do not see or experience the benefits of peace. Youth are those who have the capacity to carry a conflict into the next generation and it is therefore unlikely that sustainable peace can be built without the commitment or inclusion of youth into the process. Therefore, in order to have full engagement of youth in the peace process, difficult and sometimes unpopular decisions have to be made for the inclusion of the full range of youth.

Challenges

Former child and youth combatants in post-conflict situations face a variety of challenges to successful reintegration before they can become fully functioning members of a peaceful society. For such young people, making the transition into normal stage of the life cycle is centrally important. However, many former child soldiers in many cases do not re-enter the normal stage of the life cycle because they may have never known anything that would be considered normal. Neil Boothby director of the Program on Forced Migration and Health at Columbia University, has conducted interesting longitudinal research with child combatants in Mozambique. The first point of his findings suggest that what was really important for those young people was to find the means to re-enter a normal life cycle. This could involve, for example, accumulating enough wealth to marry, becoming a productive citizen and becoming a parent. Those were critical issues in terms of recovery, more so than any kind of psychosocial treatment provided at rehabilitation centers.

Regarding the second point, at this stage of life, identity development is critically important. For young people who have been engaged as combatants in an armed conflict, being able to enter into normal stage in the life cycle requires a process of identity transformation. This involves a transition from seeing oneself as a member of a fighting force, to seeing oneself as a member of the community. Such a transition requires significant mental adjustments and it is a huge challenge. As a result, when considering programming, it is important to think in terms of how to facilitate a process of identity transformation for young people who have been engaged in conflict.

Additionally, the risks related to HIV/AIDS must be addressed in post-conflict environments. The results are mixed in terms of the impact of armed conflict on the spread of HIV. In some contexts, conflict appears to facilitate the spread of HIV while in others its seems to retard it. However, once peace is established, the opportunity for HIV to spread increases as the population's mobility increases. In post-conflict environments, recognizing the array of immediate issues that a country must address, it is hard to keep the country focused on an issue that does not pose an immediate threat, despite the fact that it may become a serious problem in the future. However, if the spread of HIV is not addressed, there will be serious consequences.

The high level of illiteracy among youth is one of the key obstacles that is associated with youth mobilization in the field. About 85 percent of the youth are ex-combatants who missed out on their education. Their educational limitations and experience have made it even more difficult to work with them. During negotiations, for example, they are only concerned with their own interests and not those of others, and they tend to be pessimistic. How does the program overcome these challenges? How can the materials and teaching styles be made more accessible to an illiterate population? How is it possible to change these attitudes and behaviors?

Overcoming a lack of real political commitment to youth development and empowerment is a huge challenge. There is a need to engage and pressure governments to pass a youth empowerment act in order to institutionalize youth engagement structures and facilitate the empowerment of youth. In addition, the lack of direct donor support for youth development initiatives continues to pose a serious threat to effective engagement of youth in peacebuilding. Donors must be educated and a constructive dialogue established between donors and youth agencies and groups to encourage direct support. Even if youth initiatives are laudable, donors prefer to fund them through the state-related institutions. However, these institutions normally have their own agendas. Consequently, there is a need for a paradigm shift within donor support mechanisms, moving away from these often-paranoid state systems towards a more direct engagement with the youth organizations themselves so that youth engagement in peacebuilding can become more effective and constructive.

Unfortunately, several obstacles hinder such transformation. The public stereotypes and stigma associated with young people remain quite strong even after conflict has ended. Fear of young people is pervasive and it can be argued that one of the fundamental reasons why authentic youth owned, youth guided and implemented activities do not happen is due to the unwillingness to share power. Put bluntly, it is the fear or being terrified of youth. Working with youth carries the risk of unpredictability and uncertainty. While youth are far from perfect, they have the same imperfections one expects in any human being, particularly at a transitional stage in life. Sometimes youth do not make the best choices, however, in the real world so is the case with adults too. In many cases youth participation in peacebuilding processes is hindered by the denial of their basic rights and consequent struggle to survive. If youth are worried about where the next meal is coming from, then their priority is not to participate in the community as a peacebuilder. Unfortunately, there exists a relative paucity of funding for youth peacebuilding activities. Even when funding is available it tends to be short term support rather than the longer-term support needed to build significant life skills, facilitate integration into communities and build young people's basic rights occurs as a result of armed conflict initiated by adults. It is true that youth play a significant part in war, but it is also true that wars are overwhelmingly started by adults.

Even among those of us who take seriously the charge to engage youth as peace builders, and to make young people the center and the leaders of programs, our rhetoric does not live up to the reality. We need to answer the question, how can we bridge the gap between our aspirations and our actions.

Where former armed youth return to their communities, they are often not socially well received by those still living there. Are there programs that address inter-generational perceptions of young people, and how do we initiate programs that build bridges across generations? This is an area identified as one of opportunity for future efforts. Current disarmament, demobilization and reintegration (DDR)practices tend to celebrate those who were soldiers and marginalize those who never carried a gun. This effectively rewards youth for joining armed groups, or is atleast perceived as doing such by local communities. How can we avoid programs that appear to reward bad behavior? The increased attention to youth has been accompanied by a troubling blurring of categories, especially between the categories of youth and children. There is a fine line between empowering youth and undermining legal frameworks and mechanisms that protect children in areas where violence is present. Youth frequently have no separate legal definition or protection of their own, while children do. In many contexts, the rule of law in dangerously disconnected from the relevant realities of youth, reflected by the complete lack of communication between the justice/accountability sector and youth sector. In recent years, donors have been preoccupied with the creation of jobs, effectively ignoring funding efforts in other sectors. Employment rates alone are insufficient as the only indices. Better analytics are needed for these difficult-to-quantify dimensions, such as cultural change, social inclusion, networking and mentoring. Employment programs that do not adequately address young people's sense of dignity or identity could struggle to act as a conflict mitigation strategy. Donors at times fund these creative efforts, such as the CDA Listening Project3, which establishes new ways of understanding populations affected by violent conflict.

Naïve ideals, stubborn and even paternalistic belief in superiority of certain practitioner ideas, disregard for young people's ideas rather than solidarity may lead to youth disengagement in the peacebuilding efforts. Many aid workers, teachers, and others are often found to have negative and stereotypical views about young people with whom they work. These issues may have structural explanations for this dynamic among people working with youth which can be addressed. These workers are often overburdened, under-resourced and feel marginalized from higher level decision-making processes.

Engaging Youth in Peacebuilding

Today almost half of the world's population (48 percent) is under the age of 24, and of these 18 percent or more than one billion people – are defined as youth (UN, 2007). The effects of globalization and shifting economic, social and political spheres impact the daily lives of youth. Social exclusion and a lack of opportunities perpetuate youth disillusionment, which, in turn, affects their transition to adulthood. Youth are also affected by unemployment and under employment, limited education, poor governance, sexual and reproductive health issues and limited civic participation. Undeniably, these sites of structural violence are exacerbated in conflict and peacebuilding states.

Although conflict may provide an opportunity for change, if not managed correctly and peacefully, it can escalate into violence (Asian Development Bank, 2012). Moreover, in post conflict contexts, exclusion is one of the most important factors that trigger a relapse into violence (UN, 2012). Therefore, strengthening international mechanisms that specifically address youth needs is not only a demographic and democratic imperative, but also crucial in preventing conflicts from escalating into violence. Current international mechanisms do not sufficiently address the specific situation of youth peace and security.

The World Programme of Action for Youth (WPAY) review in the year 2005 is one of the most progressive documents surmising that youth are also agents of peace. With the right educational tools for crisis prevention and peacebuilding they can develop the skills needed to help prevent violent and armed conflicts (UN, 2005). However, this recognition has failed to be reflected in any enforceable programs of action or within the framework of the MDGs. The inclusion and participation of young people enhances their capabilities and affords them the opportunity to improve their lives as well as their communities. Involving youth in peacebuilding processes, as stakeholders and decision makers, allows then to gain ownership of the policies that affect and those that are around them.

Despite being both causal agents and victims of conflict and displacement, youth are one of the groups provided with the least support in conflict and post-conflict settings (Siobhan, 2011). Propelled by social exclusion, lack of opportunities, and, in the context of armed conflicts, a lack of or slow implementation of public policies that promote preparation and reconciliation, youth can become exposed and vulnerable to armed or political recruitment and exploitation.

In situations of conflict, protection mechanisms already exist for other vulnerable groups such as women and children. The UN Security Council Resolutions 1612 (2005) and 1882 (2009), formalized protection against children in conflict, including the recruitment of child soldiers, timely reporting of violations of children's rights, maiming and killing children as well as using sexual violence against them. UN Resolution 1325 on women, peace and security addresses women's equal participation in peace and reconciliation processes. In this context, it is important that youth must be seen as a different group with different potential. Ensuring the active, systematic and meaningful participation of youth in peacebuilding is a demographic and democratic imperative. By engaging youth productively in their societies and in ways that strengthen their livelihood opportunities, the vulnerability of young people can, to a large extent be addressed. A pragmatic shift in the perception of young people's role in conflict could help transform youth from being victims or agents of violence to youth as active agents of peacebuilding and positive social change. In post conflict countries, as much as physical infrastructure needs rebuilding, so does the fabric of society – relationships and importantly trust between different conflicting parties and trust between society and the government.

Young people are more open to change, innovative, concerned about long-term stability and exhibit a willingness to work hard and are in a position to facilitate the rebuilding of social capital. Yet these capacities and potentials of youth go unrecognized. Active participation in non-governmental institutions such as peace initiatives, churches and youth clubs could provide frameworks through which the relation to the former enemy can be redefined (Buckley-Zistel, 2008).

Commentators have further suggested that various opportunities exist for youth involvement at a community level. There is much to be gained from programming that works to spiral up social capital and resources that strengthen individuals, strengthen groups, increase cooperation across generation and contribute to a healthier and more vibrant community. Youth interest and engagement can increase when social capital is built alongside younger people and into the overall planning and implementation processes where young people should not be seen as a homogeneous category (UN, 2012). The terminology young peacebuilders is not mutually exclusive to youth in fragile or conflict prone states. The underpinning values of peacebuilding to promote intercultural dialogue and a culture of tolerance, diversity and equality are relevant to societies and youth groups across the globe. UNOY Peacebuilders' experience reflects the importance of human rights and peace education in the communities where they work.

Most theorizing and analyses of youth in post-conflict settings have tended to focus on the challenges and the vices they present. These observations are inclined to vies young people either as perpetrators of violence, as problems to be solved or as helpless victims of society's structures and processes. While these conceptualizations are evidently informed by realities on the ground, they have diverted the attention from the distinctive initiatives of peacebuilding and social change processes in which the youth participate. What remains lacking in the current body of literature is a focus on the positive aspects of youth engagement in post-conflict societies. The obsession with youth in peacebuilding processes, leads to unresponsive youth policies and programmes. Accordingly, this chapter embodies an emancipatory social analysis that seeks to view actors as independent agents of social reality, despite the existence of structural constraints and limitations. Scholars such as Drummond-Mundal and Cave (2007), it can be assumed that there is dissatisfaction with the mainstream perspectives of youth in post-conflict settings and which revolves around victimology. Drummond-Mundal and Cave (2007) suggest that focusing only on the vulnerabilities of young people is a limiting perspective that denies the opportunity to influence their own lives and futures, and overlooks their insights, their rights to participate and their potential to contribute to peacebuilding. Indeed, young people are active individuals, possessing assets such as resilience, curiosity, intellectual agility, innovativeness, vision of possibility and the capacity to help others (Apfel and Simon, 1996). Scholars such as Boyden and De Berry (2004), Argenti (2002), De Waal (2002), Sommers (2006) and Thorup and Kinkade (2005) now acknowledge the ability of young people to influence their fate positively. From the Soweto uprisings in South Africa to the anti-military protests in Nigeria, to the

Twitter revolution in Moldova and finally to the Arab Spring, History has demonstrated that young people are not innocent bystanders of social change, but they are innovative, creative and agentic participants in socio-economic and political processes. Youth in many parts of the world have evidently played progressive roles towards transforming situations of conflict, ultimately leading to the reconfiguration of political and social structures. In west Africa, particularly in Nigeria and Mali, youth were part of civil society coalitions that spearheaded political transformation in the 1990s, where young people resisted military regimes in their respective countries. Similarly, in the Arab Spring, youth activism had a positive transitional impact across north Africa and even extended to parts of the Middle East, such as Syria and Bahrain. In these instances, young people demonstrated that they could engage positively by challenging repressive regimes instead of being manipulated as instruments of aggression and violence. Evaluating the conflict transformation role of youth during the period of apartheid in South Africa, Drummond-Mundal and Cave (2007) concluded that young people's exercise of agency made a difference. The involvement of youth in the anti-apartheid struggle began with Soweto uprisings in 1976, and the outcome was a loosening of apartheid laws. Reynolds (1998) also observed young people's agency in the struggle against apartheid in South Africa where he found that youths chose 'to take part in the struggle against apartheid'. She stressed that they 'took profoundly serious political and moral decisions in relation to their own safety and ambitions as well as the safety and interests of their'.

Similar to that of youth being violent, there is extensive evidence of youth not only being peaceful but also of being agents of positive social change. However, this phenomenon has not been analysed by academic research. How many young people are violent and how many young people are peace-builders? Social research using quantitative and qualitative methods can help understand and answer this question. Our experiences as youth workers and educators in several contexts suggest that there are many youths who are peace-builders. They are pro-active agents in their communities, in their schools, work places, sports teams, youth groups and universities. Their stories have yet to be told. McEvoy (2001) wrote that in any conflict context, one examines, the dominant presence of the young in youth work, in community development and in the inter-ethnic dialogue and peace groups is clear. Many have direct experience of violence, conflict and have faced imprisonment themselves. They are not well paid, their projects are under-funded, often stressful and can be life threatening. Like other civil society actors, they are less visible in analysis of peace processes than key elites. One of McEvoys final propositions is that youth are the primary actors in grassroots community development/relations work. They are at the frontlines of peacebuilding. In similar fashion, Ardizzone studied the activities and motivation of youth organizations in New York city. She highlights, '*Global Kids*', an organization which conducts student-run conferences on children's rights. *Youth Force* which teaches teens how to respond when stopped by the police and Youth Peace/Roots which tries to counter the active role military recruiters take with inner-city youth by educating about alternatives to militarism. She emphasizes that young people whom she interviewed show a commitment to changing the situation of their peers and the image adults have of youth. The youth in this study have witnessed injustice and rather becoming hopeless and apathetic, or perpetrators of direct violence, have chosen to work for social change. Further, studies should be undertaken to document similar and not so visible initiatives in other geographical and cultural contexts, linked to studies of civil society strengthening. Another aspect to highlight from the conclusions of Ardizzone, is that there seems to be a parallel with the emancipating role that war sometimes have on women. Women are compelled to take on roles left vacant by men – typically soldiers – during times of war. In this case, youth take up the role of a generation of adults who are either hopeless, too comfortable to change or incapable of implementing transformation. The experiences of agencies working with youth, support this idea. The Oxfam International Youth Parliament Report, 'Highly Affected Rarely Considered', stated that the experience of the International Youth Parliament is that an increasing number of young people are rejecting violence and becoming involved in peace-building efforts at the grassroots, national and international level (Felice and Wisler, 2007).

The focus of scholarship on the potential destabilizing characteristics of youth as with the youth bulge literature, or the moral and protection issues surrounding children's involvement in conflict, as with the literature concerning child soldiers, does not provide an adequate model for understanding youth behavior in the post-conflict environment. Two recent models, however, strive first to address how varying approaches to youth programming reflect limits in current conceptions of potential youth roles, and secondly, develop a more encompassing understanding of youth behavior during post-conflict reconstruction. Frist, Kemper (2005) in her recent report, 'Youth in War-to-Peace Transitions: Approaches of International Organizations', provides a model for understanding the various approaches used to address youth issues during post-conflict reconstruction. Kemper divides the approaches into three distinct categories (i) rights based, (ii) economic and (iii) socio-political. Each approach has significant value for the reconstruction process that reflects a different understanding of children's and youths' needs and addresses a different time frame in youth development in conflict situations. For instance, as previously discussed, the rights-based approach predominantly focuses on preventive policy and views children as victims in any hostile situation that undermines their legal human rights as social actors. An economic approach moves further away from the passive role of the child. Instead, however, just as the rights-based approach is limited by its underlying principles, the economic approach is similar to the 'Greed-Grievance' model, the economic approach views youth as decision makers in the market place who make rational choices in pursuit of their best interests. As such an economic approach is driven by the potential roles of youth as either an easily exploitable resource for conflict production, or productive peacetime economic actors. The respective programming is therefore aimed at the short-term reintegration of youth into productive economic activity. However, just as the rights-based approach is limited by its underlying principles, the economic approach is limited as it reduces young men's role to simply being a resource available for manipulation and therefore inadvertently accepts the narrow view of those who exploit them and carries on myths of youth's inherent violence. As such, the more traditional child's rights and economic approaches are meant to help children and youth achieve their potential along a spectrum of potential roles in terms of both social and economic aspects. While this description of possible roles for children and youth may be descriptively accurate, it is not complete. Young people do fill these roles, however, they are not limited to them. Youth also have the potential to act as social and political agents that can contribute to the post-conflict reconstruction process. In this respect, a socio-political approach to youth programming regards youth as critical members of civil-society and understands the precarious long-term dynamic of youth as significant and active agents in the community, both as potential spoilers and as peace-builders. While socio-political youth programming is the least utilized of the three approaches, successful socio-political programmes can aim to rebuild war-torn societies through and by youth. Each of these types of programming provides a distinct value to the reconstruction process. However, Kemper notes that most often one approach is favored over another. Instead, she suggests that a holistic perspective on youth programming would include all three approaches, and recognize youth agency as social, economic and political actors (Schwartz, 2008).

In connection with youth as peace builders, the UN has laid down few principles which are listed below and are of equal importance. These guiding principles were developed by the Subgroup on Youth Participation in Peacebuilding of the United Nations Inter-Agency Network on Youth Development. The Subgroup on Youth Participation of the United Nations is Co-chaired by the United Nations Peacebuilding Support Office (PBSO) and Search for Common Ground (SFCG) and consists of members representing a member of United Nations agencies, international non-governmental organizations, academics and youth led organizations. These guiding principles have benefitted greatly from inputs from more than 1000 stake holders and organizations, including among others, communities and Boundaries, Economic Community of West African States, the Inter-Agency Network for Education in Emergencies, Inter-Peace, Mercy Corps, the Office of the Special Representative of the Secretary-General on Children and Armed Conflict, the Office of the Special Representative of the Secretary-General on Violence Against Children, PBSO, SFCG, The Unite Nations Alliance of Civilizations, United Nations Children's Fund, The United Nations Populations Fund, The United Nations Office on Sports for Development and Peace, The United Nations Programme on Youth, The United Nations Human Settlements Programme, The United Nations Volunteers Programme, The United Nations Network on Young Peace-builders, The US Agency for International Development, The Women's Refugee Commission and World Vision International. However, the ideas and opinions expressed in the principles do not necessarily reflect the views of these organizations.

Given the fact that the international focus on youth and peacebuilding has only emerged relatively recently, the mechanisms for engaging youth in peacebuilding and the content of that engagement are still in the early stages of experimentation. There are, however, two axes around which current activities and approaches that can be conceptualized. The preservation/transformation of the status quo and the internal/external orientation. Regarding the first axis, at one end of the spectrum are those programs that preserve the status quo and at the other end are those focusing on transforming it. Many, if not most programs, will fall somewhere in the middle. Programs that focus on preserving the status quo seek to maintain stability in post-conflict situations once the cessation of direct hostilities begin to emerge. From this perspective, peacebuilding programs may focus on the integration of youth within social processes and institutions in order to forestall efforts to mobilize youth to perpetuate conflict. In essence, such programs aim to remove obvious sources of political and economic discontent among young people, distinct from such programs are those that aim to transform the status quo or position youth to be agents for such a transformation. Such programs are also essential for effective peacebuilding as they aim to mobilize youth towards imagining a future without war. Transformative programs increase the amount of mental and social capital that is oriented towards embedding patterns of peace and are dedicated to that purpose. While status quo-preserving programs stabilize the present, status quo-transforming programs open up the vistas of a peaceful future.

The second axis has at one end programs emphasizing the internal lives and intimate relationships of young people, and at the other end are those focusing on the social positions of youth and their relationships with the larger society. Again, many programs fall somewhere in the middle of this spectrum. Programs that emphasize primarily on the internal life of young people, focus on engagement at the level of personal identity and worldviews, translating internal transformation into patterns of behavior conducive to peace and unity. Typically, such programs directly involve youth in reflecting upon and seeking change in, their most immediate environments, such as their personal relationships, schools and families. At the other end of the axis are those programs that focus on the place of youth within society and the kinds of social activities in which they are engaged. One focus of such programs is to provide more opportunities for youth to broaden their engagement in social and public life. Whether through undertaking projects for the betterment of society, participating in mentorship programs, forming clubs or societies that have social goals and purposes, or benefitting from programs that grant them access to social institutions (Dhanesh, 2008).

Meaningful youth engagement in post-conflict and transition settings involves long-term commitment to youth-led, adult-sponsored processes that encourages young people to express themselves to be involved in decisions that affect them and their communities and be recognized as active social actors in the communities in which they live. This requires commitment to inclusive and participatory processes that are grounded in principles of protection and participation. Civic engagement in particular is increasingly recognized as a powerful positive tool for re-engaging disenfranchised youth in society and in education systems allowing youth to build their capacities making swift transition to productive work and public life, and connection between political, social, and educational civic engagement has major connectivity in the lives of youth in both formal and non-formal education settings (Dolan, 2012). The growing literature on peace education reflects a dynamic field. Harris (2004) divides peace education into five categories (i) international education, (ii) development education, (iii) environmental education, (iv) human rights education and (v) conflict resolution education. Curricula on peace education covers a range of topics that include, history and philosophy of peace education (Reardon, 1998; Burns and Aspeslagh, 1996; Harris and Morrison, 2003), the dialectic between negative and positive peace (Galtung, 1969, 1996), gender and militarism (Reardon, 1993, 2001), conflict resolution education (Johnson and Johnson, 2006), and the formation of peaceful values in education (Boulding, 1988; Toh and Cawagas, 1991). On nurturing cultures of peace, Sommerfelt and Vambheim (2008) write that peace requires citizens to contain their aggression, exhibit cooperative behavior and resolve conflicts without violence.

REFERENCES

Apfel, R.J and Simon, B. (1996) 'Introduction', in Apfel, R.J and Simon, B. (eds.) *Minefields in Their Hearts: The Mental Health of Children in War and Communal Violence*, New Haven: Yale University Press.

Ardizzand, C. (2003) 'Generating Peace: A Study of Non-Formal Youth Organizations', *Peace and Change Journal*, 28(3): 420-445.

Argenti, N. (2002) 'Youth in Africa: A Major Resource for Change', in De Waal, A. and Argenti, N. (eds.) *Young Africa: Realizing Rights of Children and Youth, New Jersey*: Africa World Press.

Asian Development Bank (2012) '*Working Differently in Fragile and Conflict-Affected Situations – the ADB experience: A Staff Handbook*', Mandaluyong: Asian Development Bank.

Boulding, E. (1988) '*Building a Global Civic Culture: Education for an Interdependent World*', New York: Teachers College Press.

Boyden, J. and De Berry, J. (2004) 'Introduction', in Boyden, J and De Berry, J. (eds.) *Children and Youth on the Frontline: Ethnography, Armed Conflict and Displacement*, Oxford: Berghann.

Bukley-Zistel, S. (2008) '*Conflict Transformation and Social Change in Uganda: Remembering after Violence*', Basingstoke: Palgrave MacMillan.

Burns, R.J. and Aspeslagh, R. (eds.) (1996) '*Three Decades of Peace Education Around the World: An Analogy*', New York: Garland.

De Waal, A. (2002) 'Realizing Child Rights in Africa: Children Young People and Leadership', in De Waal, A and Argenti, N (eds.) *Young Africa: Realizing Rights of Children and Youth*, New Jersey: Africa World Press.

Del Felice, C. and Wisler, A. (2007) 'The Unexplored Power and Potential of Youth as Peace-Builders', *Journal of Peace, Conflict and Development*, 11(1): 27-45.

Dhanesh, R. (2008) '*Youth and Peacebuilding*', www.tc.columbia.edu/./Dhaneshyouthandpeacebuilding_22feb08.doc.

Dolan, P. (2012) 'Travelling Through Social Support and Youth Civic Action on a Journey Towards resilience', in Ungar, M. (ed) *The Social Ecology of Resilience: A Handbook of Theory and Practice*, New York: Springer.

Drummond-Mundal, L., and Cave, G. (2007) 'Young Peacebuilders: Exploring Youth Engagement with Conflict and Social Change', *Journal of Peacebuilding and Development*, 3(3): 63-76.

Emery, M. (2013) 'Social Capital and Youth Development: Toward a Typology of Program and Practices', *New Directions for Youth Development,* 2013(138): 49-59.

Galtung, J. (1969) 'Violence, Peace and Peace Research', *Journal of Peace Research*, 6(3): 167-191.

Galtung, J. (1996) '*Peace by Peaceful Means: Peace and Conflict Development and Civilization*', London: Sage Publications.

Harris, I. (2012) 'Peace Education theory', Journal of Peace Education, 1(1): 5-20.

Harris, I. and Morrison, M.L. (2003) '*Peace Education*', Jefferson: McFarland.

Johnson, D.W and Johnson, R.T. (2006) 'Peace Education for Consensual Peace: The Essential Role of Conflict Resolution', *Journal of Peace Education*, 3(2):147-174.

Kemper, Y. (2005) '*Youth in War-to-Peace Transitions*', Berghof Center for Constructive Conflict Management Report No. 10. The Netherlands: Berghof Center for Constructive Conflict Management.

McEvoy-Levy, S. (2001) '*Youth as Social and Political Agents: Issues in Post-Settlement Peacebuilding*', Kroc Institute Occasional Paper #21:OP2.

Reardon, B. (1993) 'Women and Peace: Feminist Visions of Security', New York: SUNY Press.

Reardon, B. (1998) '*Comparative Peace Education: Educating for Global Responsibility*', New York: Teachers College Press.

Reardon, B. (2001) *Education for a Culture of Peace in a Gender Perspective*, Paris: UNESCO.

Reynolds, P. (1998) 'Activism, Politics and the Punishment of Children', in Van Buren, G. (ed) *Childhood Abused*, Hampshire: Ashgate.

Schwartz, S. (2008) '*The Dynamic Role of Youth in Post-Conflict Reconstruction: Lessons from Mozambique, The Democratic Republic of Congo and Kosovo*', Unpublished Theses, Wesleyan University. www.un.org/./peacebuilding/.Guiding per cent 20principles per cent 20Youth per cent 20pa.cached.

Siobhan, M. (2011) 'Children, Youth and Peacebuilding', in Matyok, T., Senehi, J., Byrne, S (eds.) *Critical Issues in Peace and Conflict Studies: Theory, Practice and Pedagogy*, London: Lexington Books.

Sommerfelt, O.H. and Vambheim, V. (2008) 'The Dream of the Good – A Peace Education Project Exploring the Potential to Educate for Peace at the Individual Level', *Journal of Peace Education*, 5(1): 79-96.

Sommers, M. (2006) '*Youth and Conflict: A Brief Review of the Available Literature*', Washington, D.C: USAID.

Thorup, C.I. and Kinkade, S. (2005) '*What Works in Youth Engagement in the Balkans*', Baltimore: International Youth Foundation.

Toh, S.H. and Cawagas, V.F. (1991) '*Peaceful Theory and Practice in Value Education*', Quezon City: Pheonix.

United Nations (2007) '*United Nations General Assembly Resolution 50/81 of 13 March 1996*', *The World Programme of Action for Youth in the Year 2000 and Beyond*, A/Res/50/81.

United Nations Department of Economic and Social Affairs (2005) '*World Youth Report 2005: Young People Today and in 2015*', New York: United Nations.

United Nations Peacebuilding Support Office (2012) '*Informal Brainstorming Consultation on Youth Participation in Peacebuilding Report*', New York: United Nations.

United Nations Secretary General (2012) '*Peacebuilding in the Aftermath of Conflict*', 1/67/499.

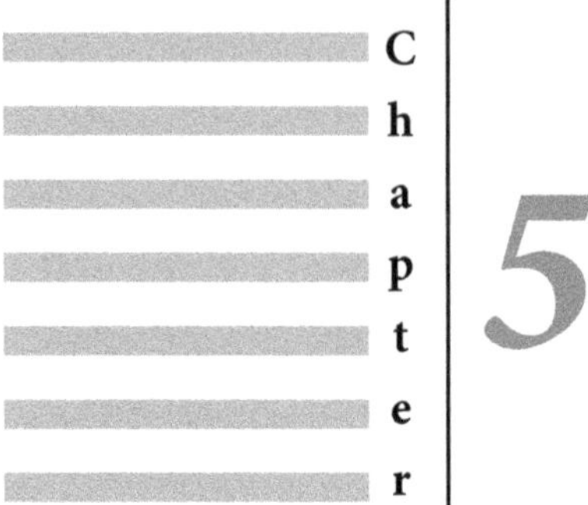

Conflict in North East India

The idea of north east India and the conflicts in the region can never be easily comprehended. Since the independence of India, many of the conflicts in the region have lingered for more than half a century and have become more complex (Phanjoubam, 2016). India's north east, consists of eight states namely (i) Arunachal Pradesh, (ii) Assam, (iii) Manipur, (iv) Meghalaya (v) Mizoram, (vi) Nagaland (vii) Sikkim and (viii) Tripura. At the time of independence of India, the north east region consisted of the state of Assam, and the union territories of Manipur and Tripura. In the year 1971, following the North Eastern Areas (Re-organization) Act, the north eastern region became a significant administrative concept with the formation of the North Eastern Regional Council (NERC). Sikkim became a protectorate of India in 1947 and became a full state of the Indian Union in the year 1975 (Singh, 1987). The north east region is home to 475 ethnic groups and sub-groups and speak approximately 405 dialects (Bhaumik, 2009). In the context of the colonial background, a cursory examination demonstrates the ubiquity and its effects, which is seen even after more than five decades of the dismantling of the colonial empire that has played out in a tragic fashion. The colonial legacy can be identified in two contexts (i) collective historical trauma and (ii) a causal variable that continues to have an impact on the ways in which the outlook of the states has been, post decolonization (Miller, 2013). In the year 1757, Bengal was conquered by the British East India Company and to the north east of Bengal was the Brahmaputra river valley which was predominantly inhabited by the Assamese. The British East India Company brought this region under its control in the year 1826 after the war with Burma (now called Myanmar). After the British took control of the region most of the hill areas and some plains were designated as tribal areas and was closed for immigration and kept under the control of administrative regimes. In addition, Bengal and Assam were interspersed with the princely states of Bhutan, Nepal, Manipur, Sikkim,

Tripura, Cooch Behar and the Khasi States. The British also claimed certain parts of the Himalayas but did not venture into the more hilly and mountainous areas. These arrangements of the colonial map of the region reflected and reinforced tremendous political and ethno-linguistic heterogeneity (Lacina, 2009). Colonial policies that were made during the years 1874 and 1934, segregated the region where tribal populations were administered under the category of non-regulated, backward or excluded areas. Such policies prevented outsiders from entering the region except those who obtained special permission from the government under the Inner Line Regulation of 1873. Further, an extension of this regulation to almost all hill regions resulted in the creation of a frontier within a frontier that highlighted the political and cultural rift between the people of the plains and the tribals living in the hill areas. These factors created a condition where the tribal areas were administered differently vis-à-vis other regions of the country. The Government of India Act of 1935, continued with this policy of exclusion due to which the overriding effects were twofold. While there were socio-political developments taking place elsewhere in the country even during the British period, north east India continued to be excluded from such initiatives resulted in the failure to integrate many of the tribal communities. These exclusionary policies resulted in unequal and unbalanced development of the region that was present during the British period, continued to exist even in the post-independence period which created conditions for ethnic conflicts to emerge in the region. In response to the emergence of these problems the Government of India created autonomous districts and regions that were often identified with tribal affiliations through constitutional measures such as the Sixth Schedule. Many of these regions subsequently became full states and demands for such similar arrangements for homelands to be created were made by the tribes living in the region (Upadhyay, 2006),

Geographically India is connected with the north east of India with a narrow strip of land known as the Siliguri corridor or otherwise known as the chicken neck. Any historical analysis of this region would reveal that it has been host to a multitude of ethnic groups whose characteristics and value systems would direct the social, economic and political interactions which are important to them. Crucial to this is a clear understanding of these issues that would help those who are tasked with solving the region's problems. People who are living in the region are migrants who have settled through a process of perpetual migration. These migrations were ethnic in nature and can be characterized as belonging to the mongoloid groups belonging to greater Tibet, Mongolia and China, the Mon Khmer groups Aryans, Negritos and Dravidians from the west are among the rest who had migrated to this region (Mukherjee, 2005).

North east region of India known as north east India had never been historically part of the Indian union. This region was forcefully annexed by the British (Bhaumik, 1998). Following the independence of India from the colonial rule in 1947, this region has been vociferously demanding secession from the Indian union (Innoue, 2005; Vadlamannati, 2011). It is in this background that most scholars on north east India study the region as a theater of insurgency and counter-insurgency. Although there is this common assumption among policy makers and social scientist who want to examine the problems in north east India, it must however, be taken into account that the problems of the region are complex and different from each other (Misra, 2000; Freddy, 2017).

Post-independent India has been witness to a host of separatist and insurgent movements. However, India has not lost any of its territory despite the multiplicity and intensity of these insurgent movements that have been demanding secession. Although new to the Indian state, it required an untested government and military that had to adapt to a form of political warfare in which they had little or no experience at all (Ladwig, 2009). The objective of this chapter is to present to the reader a brief description of the conflict in north east India that would help in understanding the complexity and provide a direction while reading the following chapters in the book. It does not engage with a critical analysis of the conflicts, but provides an overview of the conflicts that have been present in north east India. It is in this context that it would be noteworthy to identify three important insurgencies that are significant and began in the aftermath of the independence of India after 1947. Insurgencies in Assam, Nagaland and Manipur have been the most significant in the aftermath of the independence of India and by the 1970s the region had become host to a multitude of insurgent movements which carried the intent of achieving secession or autonomy more particularly in the context of the Nagas and the Assamese. In the early 1990s all states, with the exception of Sikkim, plunged into some sort of insurgent activity that forced the Indian government to recognize these movements as low-intensity conflicts. Much of the conflict in the region has been attributed to the failure of the Indian government which failed to recognize the ethnic and cultural specificities during the formation of the states and the delineation of the states themselves that led to the discontentment with the Indian state and assertion of the north eastern group identity (Das, 2007). The Indian government's response to these developments was unfortunately the use of force and the government resorted to the implementation of the Armed Forces Special Forces Act (AFSPA) 1958 to contain the situation. The AFSPA has been viewed as a draconian act that enables security forces to launch counter insurgency operations with impunity. This has resulted in massive violations of human rights and has increased resentment among the people living in the region against the Indian state. Although there has been a decrease in insurgent activity in the region it is interesting to note that they are still active. Besides insurgent movements in north east India, there has been the presence of inter-ethnic rivalry that has been over resources and territory. In addition to insurgent movements in the region these inter-ethnic conflicts have pulled the government in different directions and it has been a difficult task for the Indian government to provide solutions. As a result, protests, strikes, public curfews, economic blockades by public organizations who act on the direction of insurgent groups have become common in the region (Sandham, 2005; Shimray, 2004). However, interestingly the people in the region who have never been appreciative of the actions of the Indian state have now become dependent on it for solutions over issues relating to territorial integration such as greater Nagaland, *etc.*

Among the longstanding conflicts between the state insurgent groups in the north east, the Naga conflict has been the first. The Naga conflict which began with the demand for autonomy for the protection of the 'Naga Way of Life' soon turned into insurgency that was led by the Naga National Council (NNC) in the early 1950s (Misra, 2000). The self-perception of the Naga national identity was manifested with the emergence of the NNC which spearheaded the separatist movement of the Nagas and continues to identify its workers as guerillas (Bhaumik, 2009). In terms of military might, the Naga socialist Council of Nagaland – Isak and Muivah (NSCN-IM) has been the most formidable in the region. In the states of Assam and Manipur organizations such as the United Liberation Front of Asom (ULFA), Bodo Liberation Tiger Force (BLTF), People's Revolutionary Party of Kangleipak (PREPAK) and the Kuki National Army (KNA) have been at the forefront of insurgent activity and have been the most active in their respective states. While these above mentioned groups are prominent, it is extraordinarily interesting to note the presence of numerous militant organizations in the region (Baruah, 2002).

The right to self-determination was invoked by the Nagas as they possessed a distinct identity and unique history to justify their demand for secession from the Indian union. Although the initial demand for secession was made through peaceful protests, the Nagas resorted to armed struggle to achieve independence from the Indian union. Contrary to the expectations of the Indian leadership the Naga insurgent movement could be suppressed through military force, it has remained active even after five decades since it began. This had resulted in the change of policy by the government of India since the 1990s which has tried to establish peace through negotiations and peace settlements. These initiatives which were undertaken by the government of India led to the cease-fire agreements between the NSCN-IM in the year 1997 and the National Socialist Council of Nagalim-Kaphlang (NSCN-K) in the year 2001. The most recent development of these negotiations was the Naga Peace Framework which was agreed upon by the government of India and the NSCN-IM in the year 2015 (Srikanth and Thomas, 2005; Goswami, 2015). However, despite these developments, there remains a host of problems that relate to the integration of greater Nagalim and the inter-ethnic group rivalries between the Nagas and other ethnic communities living in neighboring states that pose challenges towards achieving peace in the region (Srikanth and Thomas 2005). The Naga demand for the integration of Naga inhabited areas has been a sticking point in negotiations in which Assam, Arunachal Pradesh and Manipur have categorically rejected the territorial division of their states with regard to greater Nagalim.

As noted earlier, the process by which the Indian state was put together after its independence, often through force, it provided the basis for resentment for those historically independent nations before the British rule in the region that united all these areas for their administrative convenience. In the case of the north eastern region, during the amalgamation of the region, critical issues of importance such as power sharing and governance, demarcation of federating units and economic development were left unaddressed by the Indian government to a large extent. Additionally, the exclusionary policies adopted by the Indian state subsequently resulted in conflict within various ethnic communities and against the Indian union which seems to have become a norm in the post-colonial period of India and more particularly in the context of north east India. In particular, Assam witnessed numerous conflicts of varying intensity which have ranged from mass civil disobedience that emerged as a byproduct of long standing grievances against the state to armed insurgency which favored secession from India and communal violence, resulted in genocide and ethnic cleansing (Goswami, 2014). Conflicts arose in Assam following the Indian Government's failure and reluctance to prevent the influx of refugees into the state after the formation of Bangladesh in the year 1971. The issue of Bangladeshi refugees which was illegal in most cases was a matter of concern for the people of Assam as they feared demographic swamping and the loss of Assamese identity. The people of Assam raised their concerns and protested against such reluctance shown by the Indian government on refugee influx from Bangladesh through mass civil disobedience movement that began in 1979 and continued till the year 1985 when the Assam Accord was signed. Simultaneously with the mass civil disobedience movement in Assam, emerged militant outfits which carried the agenda of secessionism among which the ULFA became the most prominent and powerful insurgent organization in the state of Assam. The ULFA's primary objective is to achieve an independent federal Assam in which all ethnic communities can co-exist vis-à-vis the Assam movement which had a different impact on the people of the state where every ethnic community had begun claiming their own ethnic identity based on which they too began demanding separate ethnic homelands (ibid). Although in a very rudimentary form, the Assamese had the idea of secessionism which was seen in various important regional movements such as the language movement (1960), refinery movement (1967) the movement for the medium of instruction (1972) and the anti-foreigner movement (1979-1985) (Mahanta, 2013). The idea of secession is deeply rooted in the history of Assam. During the framing of the Constitution of India, a section of the elite Assamese population expressed their sentiment towards secessionism as they feared that within the framework of the Indian Constitution, the legitimate interest of the Assamese would not be protected (Phukan, 1996).

As already noted, the Assam movement seemed to have been the catalyst and also one that laid the foundation for the growth of the independence of Assam which was led by the ULFA. Although many writers have expressed their doubts over the democratic nature of the Assam movement, it still remains the single most popular movement in the post independent history of India. A strong emotional attachment towards their culture and identity was a strong crowd puller for this movement (Mahanta, 2013). The assimilation of the Assamese towards the idea of a separate state can also be attributed to the actions of the Indian state during the 1983 state elections which police firing led to the deaths of more than one hundred and thirty people. Even though the Assam Accord was signed in the year 1985, displeasure, discontent and the confidence of the people of Assam was shaken (Misra, 2000). In all this the Indian government was well aware of the growing idea of secessionism in the state of Assam. The Home Minister on March 14, 1983, said in Lok Sabha, that posters and leaflets with slogans have appeared from the beginning of the movement which clearly indicate the mindset of the people who have been involved in the movement. Some of the slogans that were raised were (i) 'we shall have our country with the blood of martyrs', (ii) 'when Assam will be free', (iii) 'India has no right to rule Assam', (iv) Assam region should think of an independent United State of Assam after separation from India', (v) 'Indian Dogs leave Assam'. These statements along with the many other leaflets were distributed to the people of Assam.

The All Assam Students Union (AASU) and the All Assam Gana Sangram Parishad (AAGSP) although denied any association with these activities, the never condemned such initiatives (Sethi, 1983; Trivedi, 1985).

The ULFA and the NSCN-IM over the years have developed links with each other and Paresh Baruah leader of the ULFA reportedly met with Osama Bin Laden in Karachi during a visit to Pakistan in 1996. Earlier in the 1990s the ULFA had also developed some links with some officers of the Royal Bhutan Army and also with the ISI. The ULFA trains its cadre in a number of training camps that are both in India and also in Bangladesh. In Assam it is not only the ULFA that is operating as an insurgent organization but there are other militant groups that are ethnically motivated such as the National Democratic Front of Bodoland (NDFB) and the Bodo Liberation Tigers (BLT). The NDFB has similar demands as that of the ULFA. However, there are contrasting ideological differences between them. The ULFA has the objective of creating a separate state for all ethnic communities living in Assam while the NDFB was formed with the motive of creating a separate autonomous region for the Bodos in Assam. Later the NDFB's objective turned to the demand for the separate state for the Bodos called Bodoland. The NDFB was led by Ranjan Daimary. The NDFB claimed that the Assam Accord signed by the Assam Gana Parishad did not address the needs of the indigenous people living in Assam. the NDFB targeted migrant Muslims, Bangldeshis, Nepalis and Santhals. The NDFB carried out ethnic cleansing activities in the Bodo region through arson, bombings, extortion, kidnappings and looting of non-Bodos (Baruah, 2006). The Bodo LiberationTigers also known as the Bodo Liberations Tiger Force (BLTF) was established in the year 1996 whose objective was to create an autonomous district council along the southern banks of the Brahmaputra and the inclusion of the Bodos of the Karbi Anglong district into the sixth schedule of the Indian Constitution (ibid). on December 6, 2003, members of the BLT surrendered to the Indian government that marked the end to the insurgent movement. On December 7, 2003, the interim executive council consisting of twelve members formed the Bodoland Territorial Council (BTC) in Kokrajhar district of Assam through which demands of the BLT were met and a Memorandum of Settlement (MOS) for the creation of BTC was made (ibid).

In terms of conflict or insurgency in the north eastern region of India, the state of Manipur has witnessed significant levels of violence that have ranged from insurgent violence, terrorist attacks, extortion, kidnappings for ransom, bandhs, economic blockades, *etc.* The violence and unrest in the state of Manipur as it exists today began in the early 1970s. However, the genesis of the conflict can be traced back to the closing years of the 1930s when Manipur was not part of the British Empire and remained a native state. The Maharaja (King) of Manipur who was cruel and avaricious created wrong economic policies which resulted in the great famine in the state. It was in this situation the Manipuri Women's Uprising (Nupi Lan) of 1939 was a watershed event that seems to have laid the foundations for the present uprisings and insurgent movements in Manipur. Irabot Singh, the son-in-law of the Maharaja joined the women's uprising with the intent of overthrowing the king from his throne to make Manipur an independent republic. In the words of Homen Borgohain, it was in the soil of Manipur that the seeds of secession were sown for the Mongoloid race who inhabited the north eastern region and the man who was behind this was Irabot Singh (Borgohain, 1982).

After the state of Manipur was merged with the Indian Union the people of Manipur desired that they be treated equally with other Indians despite their distinct cultural heritage and ancient civilization. However, the Indian state remained insensitive to the desires and demands of the Manipuris (ibid). Among the insurgent groups present and active in the state of Manipur are the Meitei led United National Liberation Front (UNLF) that has been at the forefront and the oldest insurgent group in the state. The group was established in the year 1964 and was led by Areambam Samendra Singh whose objective was to establish an independent and socialist Manipur through armed struggle. In the 1990s the groups also began social reformation activities that were directed against rampant alcoholism, drugs, gambling and drug peddling. The 1990s also saw the formation of the Pan Mongoloid Coalition called the Indo-Burma Revolutionary Front (IBRF) that was started by the UNLF, NSCN-K and the Kuki National Army (KNA). The UNLF has been in conflict with the NSCN-IM which is accused of having created conflict in the state of Manipur because of its demand for the inclusion of four districts of Manipur (Mao, Senapati, Ukhrul and Tamenglong) into the state of Nagaland to which the UNLF is critically opposed to. The UNLF has also accused the NSCN-IM for having instigated the first Naga-Kuki bloodbath of 1993 (Barua, 2006).

Other insurgent groups in Manipur are the Kuki National Front (KNF) and the People's Liberation Army (PLA). The KNF has been active in terms of the demands of the Kuki groups. The PLA has been working in terms of bringing all major ethnic groups such as the Meiteis, Nagas and Kukis together to fight for an independent state of Manipur.

The conflicts in north east India are complex and diverse in nature since each state has its own demands and complexities. However, the response of the Indian state has been the use of brutal force in order to control and suppress the violence in the region (Prabhakara, 2007). India's approach towards establishing peace in the region first been on containing violence committed by insurgent groups through the use of extensive military force. However, in terms of establishing peace in the region seems to be a combination of India's diplomacy with Bhutan and Bangladesh coupled with the grassroot empowerment of communities in north east and state intelligence and policing that has resulted in some success. There has also been shifts in the policy of the government of India in negotiating peace in north east eversince the NDFB-S killed 70 civilians in Assam during the year 2014. The government of India has now decided firmly that it will not engage in peace talks with any outfit that involves in the killing of civilians and would treat such groups as terrorist entities. Such hard resolves by the government of India has resulted in the beginning of full scale operations against the NDFB-S in which senior leaders were arrested and more than 30 cares were also arrested (Hussain, 2015; Hussain, 2015a).

It is interesting to note that the conflict in north east India has sustained itself for over five decades since the independence of India. The Indian Government's approach towards the conflict and insurgency seems to be containment of violence which is seemingly equated with peace in the region. However, lack of development in the region, failure to address the needs of the people in this regard has enabled these movements to sustain themselves against the odds that have been placed against them.

REFERENCES

Baruah, S. (2002) 'Gulliver's Troubles: State and Militants in North East India', *Economic and Political Weekly*, 37(41): 4178-4182.

Bhaumik, S. (1998) 'North East India: Evolution of Post-Colonial Region', in Chatterjee, P. (ed) *Wages of Freedom*, New Delhi: Oxford University Press.

Bhaumik, S. (2009) '*Troubled Periphery: Crisis of India's North East*', New Delhi: Sage Publications.

Borgohain, H. (1982) 'Manipur: Anatomy of Despair', *Economic and Political Weekly*, 46(47): 1858-1860.

Das, B. (2010) '*KNLF Rebels Surrender in Assam*', http://in.reuters.com/article/idINIndia-46094620100211. Accessed: 17 June 2019.

Das, S.K. (2007) '*Conflict and Peace in North East: The Role of Civil Society*', Policy Studies No.42. Washington D.C.: East-West Centre.

Freddy, H.J. (2016) 'Engaging Youth: Challenges and Opportunities Towards Building Peace in North East India', *Man and Society: A Journal of North East Studies*, 13(1): 106-120.

Goswami, U. (2014) '*Conflict and Reconciliation: The Politics of Ethnicity in Assam*', New Delhi: Routledge.

Hussain, W. (2008) '*The ULFA Mutiny*', *Outlook India*', July 03, 2008, http://outlookindia.com/article.aspx/237821. Accessed: 17 June 2019.

Innoue, K. (2005) '*Integration of North East: The State Formation Process, in Sub-Regional Relations in Eastern South Asia: With Special Focus on India's North Eastern Regions*' Study No. 133. Tokyo: Institute for Developing Economies.

Lacina, B. (2009) 'The Problem of Political Stability in North East India: Local Ethnic Autocracy and the Rule of Law', *Asian Survey*, 49(6): 998-1020.

Ladwig, W. (2009) '*Managing Separatist Insurgencies: Insights from North Eastern India*', Paper Presented at the International Studies Association Conference, New York, 15-18 February 2009.

Mahanta, N.G. (2013) '*Confronting the State: The ULFA's Quest for Sovereignty*', New Delhi: Sage Publications.

Mathur, A. (2011) 'A Winning Strategy for India's North East', *Jindal Journal of International Affairs*, 1(1): 269-298.

Mazumdar, A. (2005) 'Bhutan's Military Action Against Indian Insurgents', *Asian Survey*, 45(4): 566-580.

Miller, M. (2013) '*Studies in Asian Security: Wronged by Empire: Post-Imperial Ideology and Foreign Policy in India and China*', California: Stanford University Press.

Misra, U. (2000) '*The Periphery Strikes Back: Challenges to the Nation-State in Assam and Nagaland*', Shimla: Indian Institute of Advanced Study.

Mukherjee, J.R. (2005) '*Insiders Experience of Insurgency in India's North East*', London: Anthem Press.

Phanjoubam, P. (2016) '*The North East Question: Conflict and Frontiers*', New York: Routledge.

Phukan, G. (1996) '*Politics of Regionalism in North East India*', New Delhi: Spectrum Publications.

Prabhakara, M.S. (2007) 'Separatist Movements in North East India: Rhetoric and Reality', *Economic and Political Weekly*, 42(9): 728-730.

Sandhan, O. (2005) '*Who Rules Manipur's Streets?*', Article No, 1839. New Delhi: Institute of Peace and Conflict Studies.

Sethi, P.C. (1983) '*Home Minister of India Statement in Lok Sabha, 14 March 1983', in Documents of Assam, Part B*, edited by Trivedi, V.R. New Delhi: Omsons Publications.

Shimray, U.A. (2004) 'Socio-Political Unrest in the Region Called North East India', *Economic and Political Weekly*, 39(42): 4737-4643.

Singh, B.P. (1987) 'North East India: Demography, Culture and Identity Crisis', *Modern Asian Studies*, 21(2): 257-282.

Srikanth, H. and Thomas, C.J. (2005) 'Naga Resistance Movement and the Peace Process in North East India', *Peace and Democracy in South Asia*, 1(2): 57-87.

Updhyay, A. (2006) 'Terrorism in North East: Linkages and Implications', *Economic and Political Weekly*, 41(48): 4993-4999.

Vadlamannati, K.C. (2011) "Why Indian Men Rebel? Explaining Armed Rebellion in the North Eastern States of India 1970-2007', *Journal of Peace Research*, 48(5): 605-619.

Vidisha, B. (2006) '*Terrorism in India*', Huntsville: Sam Houston State University.

Peace Processes in North East India

Peace building in recent times has become a term that is commonly used in the context of conflict ridden societies which broadly means the activities undertaken to prevent, resolve violent or potentially violent conflict (Goodhand and Hulme, 1999: 15). Peace building often seems to be an umbrella term which includes a number of activities, levels and actors involved in the process (Llamazares, 2005: 3). With the end of the Cold War and the rise of intra-state wars/conflicts, the international community was involved in developing new forms of engagement to ensure development. During the time of its first usage, peace building was not seen as a concept that sought to transform societies in or emerging from conflict, but one which was engaged in maintaining stability (Denskus, 2007: 657). In the words of Featherston, 'if conflict is caused, enabled and reproduced by particular social structures and institutions which favour a dominant group, we cannot hope to alleviate or remove those causes, without altering those structures. Then peacekeeping becomes another aspect of a system which only seeks stability within the confines of that system, a system which already made the war possible' (Featherston, 2000: 196).

Throughout history it is evident that it has never been easy to mitigate armed conflicts and promote peace. Since its appearance in the mid-nineteenth century through the initiatives of various international peace-supporting institutions, it has emerged in various ways which sought more effective methods to build lasting peace. Therefore, there is a general agreement on the broad definition of peace processes as a wide range of international, national and local efforts to minimise, eliminate and transform violent conflicts (Ozerdem and Lee, 2015). It must be noted at this juncture that sustained peace is not a naturally occurring phenomenon Space is required between the two terms. It requires conscious and continuous attention which must be constructed and renewed according to the needs which might arise from time to time. States which have experienced violent internal conflicts have had to face more daunting challenges in their attempts to build peace (Jenkins, 2013: 1). While civil wars often denote the failure of legitimate state authority, sustainable peace is directed towards its reconstruction. Peace building in general is what happens and needs to be done in between the two conflicting parties (Doyle and Sambanis, 2000: 779). In this context, it is interesting to examine the efforts undertaken by the Indian state which has been engaged in bringing sustainable peace in the north eastern region for over five decades since 1947. In this essay, the focus of the peace building processes which have been undertaken over the past ten years are examined.

Since the time of its independence in 1947, India has been witness to a host of separatist and insurgent movements. However, interestingly India has not lost any of its territory despite the multitude of such secessionist movements. These insurgencies were not only new to the Republic of India, they were the first of its kind the country had to face and some of the very fierce separatist insurgent movements south Asia has ever experienced. This primarily required an untested government and military which had to adapt itself to a form of political warfare in which they had little experience (Ladwig: 2009).

Popularly known as the north eastern region or the north east, this region in India is comprised of eight states namely Assam, Arunachal Pradesh, Manipur, Meghalaya, Mizoram, Nagaland, Sikkim and Tripura. In the year 1947, Sikkim became a protectorate of India and made into a full state in the year 1975. The north east region is connected with the rest of India by the Siliguri corridor of West Bengal. Rich in natural resources, covered with dense forests, this region receives the highest rainfall in the country, with large and small rivers flowing through the land and is a cache of different flora and fauna. Diverse in cultures, customs, languages and traditions, northeast India is home to multifarious social and ethnic groups. Stemming from ideas of democracy and the practice of discussion prevailing among tribes, people of north east India have high self esteem which makes them averse to accepting the diktat of the imagined 'others'. Although such descriptions of northeast India seem to frame a picture perfect, this region has been experiencing ethnic conflicts, low productivity and market access, poor governance and lack of infrastructure. Government's inability to address problems caused by remoteness, seclusion, backwardness, have provided fertile grounds for breeding armed insurgencies and various other conflicts in the region related to identity and ethnicity. What makes the region distinct from the rest of India, are the assertions of various ethnic identities and the attitude of the state in containing ethnic extremism.

Northeast India is most commonly studied by scholars as a theatre of insurgency and counter insurgency by many scholars. A high degree of mutual alienation marks the relationship between the population of northeast India and Indians on the other side of the narrow corridor that connects the region to the rest of the country. Integration without consent, colonial attitudes, nativism, legal and illegal migration relative deprivation, cultural nationalism and irredentism have sparked violent conflict in the region for over five decades (Freddy, 2016: 107-08).

In order to provide a fair evaluation of the peace building efforts in north east India it is important to admit that the preconditions to such initiatives were rather difficult given the multiplicity of problems which were and are present in the region. The complexity of problems which the region faces are considerable and it has experienced over five decades of conflict which has left the region's human resources and infrastructure clearly underdeveloped vis-à-vis the rest of the country.

Peace Building in India's North-East

The Indian political premise of peace building seems to be based on the following (a) the state is strong and administrative and police measures work, (b) peace building measures must not be initiated unless a suitable occasion arises, (c) through a mix of strong responses even repressive and almost deliberative delays in addressing demands the insurgent groups are sought to be suppressed through military force/might, (d) the assumption that limited autonomy is the best solution (Ghosh: 2013).

While most study north east India as a theatre of conflict or insurgency or ethnic movements and counter insurgency, this region receives attention considerably only when there is an outbreak of large scale violence or when there are efforts taken to sign peace accords (Rajagopalan, 2008:1). Efforts towards building peace in north eastern region of India it would be interesting to note that thirteen peace accords were signed between 1947 and 2005. For example, in 1947, 1960 and 1975 three peace accords were signed between Naga rebel groups. The 1947 Accord also known as The Naga – Akbar Hydari Accord (SATP: 2001), which promised the Nagas autonomy to a certain extent. In 1960, the sixteen point agreement was agreed by which the Nagas gained statehood (Bose: 2013). With cracks developing in the Federal Government of Nagaland and within the Naga nationalists and the initiative of the Church (Shimray, 2005: 85), the Shillong Accord was Signed in 1975 where a few underground groups of the Nagas surrendered. However, many of the Naga nationalists including Th, Muivah and Isak Chisi Swu were not happy with the Accord and expressed their discontent (Rajagopalan, 2008: 2).

Date	*Accord*	*Parties to the Accord*	*Features*
1947	Naga-Akbar Hydari Accord	Akbar Hydari Governor of Assam and Naga National Council	Nagas were given a measure of autonomy for ten years, but the terms of autonomy were not classified, and conflict continued
1960	Sixteen-Point Agreement	Government of India and Naga Peoples Convention	Created the state of Nagaland. The Naga National Council which was underground refused to recognize the agreement
1975	Shillong Accord	Governor L.P. Singh of Nagaland and the Underground Organisations	The underground organisations surrendered, but the Accord was seen as not beneficial to the Nagas.
1985	Assam Accord	AASU and AGSP representatives, Union Home Secretary and Chief Secretary of Assam	Immigrants who entered Assam between January 1, 1996 and March 24, 1971, were to be registered under Foreigners Act, their names deled from the voters list for ten years and restored thereafter. Those who had previously been departed but reentered would be expelled. All who immigrated to Assam state after 25, 1971, would be deported under IMDT ACT, 83.

Date	*Accord*	*Parties to the Accord*	*Features*
1986	Memorandum of Understanding	Government of India and Mizo National Front leader Laldenga	Mizo National Front and Affiliates give up violence and demands for secession. They also give up links with TTNU, PLA and other armed groups. Statehood is granted.
1988	Memorandum of Understanding	Tripura National Volunteers and Government of India	Restoration of tribal lands and prevention of further alienation. Reorganization of the Tripura Tribal Areas Autonomous District Councils (TTAADC) to include tribal areas and exclude non-tribal areas. State boundaries were also secured.
1988	Darjeeling Hill Accord	Gorkhaland National Liberation Front, West Bengal State and Government of India	Sikkim is separated. Statehood demand is dropped but not removed completely from its political discourse. Ghising keeps raising issues about states of Darjeeling Vis-à-vis Nepal and about administrative level of DGHC. Also since the DGHC is not coterminous with the district, a diarchy exists in Darjeeling.
1993	Memorandum of Understanding	All Bodo Students Union, Assam State Government in the presence of Government of India Ministers and Chief Minister	Surrender and rehabilitation of ABSU cadres in return for establishment of Bodoland autonomous Council.
1993	Agartala Agreement	Tripura Government and the All Tripura Tiger Force	Renews commitment to recognizing the TTAADC and provides cultural safeguards for Tripuri's
1994	Memorandum of Settlement	Mizoram State Government and Hmar People's Convention	Chinlung Hills Development Council established
1995	Memorandum of Understanding	Assam State Government and representatives of community organisations of the Rabhas, Karbis, Tiwas and Mishings	(i) Karbi Anglong District Council becomes Karbi Anglong Autonomous Council. (ii) Rabha – Hasang Autonomous Council, Tiwa Autonomous Council and Mishing Autonomous Council, which were not territorial were established
2003	Bodo Territorial Council Memorandum	Government of India, Assam Government and Bodo Liberation Tigers	Bodo Territorial Council established, plus cultural provisions
2005	Memorandum of Understanding	Mizoram State Government and BRU National Liberation Front	Government agrees to repatriate displaced Reangs in Tripura. The Mizoram Scheduled Tribes list will now list Reangs as BRUs

Source: Swarna Rajagopalan: 2008.

The Assam state government had also entered into many agreements with the Bodo separatists and other ethnic groups as well to establish territorial and non-territorial representative arrangements aimed at providing autonomy to a certain extent. In all, the persistent problem with the peace accords signed in north east India was that each set of negotiations included only a small subset of stakeholders who were active nationalists in the region. A lack of clarity also beleaguers the autonomous Darjeeling Hill Council established by the 1988 accord between the Government of India, the Government of West Bengal State, and the Gorkhaland National Liberation Front; it is not clear to which administrative tier of the Indian Union the new district belongs. Accords signed in Tripura State appear to have made no difference to the escalating violence because militant groups there have been functioning more like transborder criminal organizations than ideologically or politically motivated groups with whom there is a starting point or a core set of issues for negotiation. In spite of the fact that illegal immigration, nativism, and citizenship are contentious almost everywhere in Northeast India, only the 1985 Assam Accord has addressed these questions, yet the provisions of that accord have not proven viable or easy to implement.

The Mizo Accord of 1986 is considered uniquely successful. One important reason for its success has been the ethnically inclusive mobilization of the Mizo National Front around a regional rather than ethnic identity. The accord however, addressed a single group's concerns over others, and cracks in this unified front have necessitated further accord making with the Bru and Hmar ethnic groups (*ibid*).

The Indian approach towards establishing peace in the region has first been the containing of violence committed by the insurgent groups by the extensive use of military force. But what made the turn around of insurgent movements inch towards peace, seems to be a combination of India's diplomacy with Bhutan and Bangladesh, the grassroot empowerment of communities in north east and state intelligence and policing which has proved to be successful to some extent. Shifts in the policy of the government in negotiating peace in north east has been noticed since the NDFB-S killed about 70 civilians in Assam in December 2014. The government of India has now decided that it will not engage in peace talk with any outfit which involves in killing civilians and would treat such groups as terrorists. Such hard resolves from the Centre's end has resulted in the beginning of full scale operations against the NDFB-S in which senior leaders and arrests of more than 30 cadres and also some commanders were also achieved (Hussain: 2015; Hussain: 2015a).

Engaging in peace talks with insurgent groups in north east India has been a frequent mode used by the Government of India to resolve conflicts in the region. Virtually almost all insurgent groups or factions in the region has entered into some form of truce with the government of India. Some of the major north east insurgent outfits engaged in talks are United Liberation Front of Asom (ULFA), National Democratic Front of Bodoland Progressive (NDFBP), National Democratic Front of Bodoland Ranjan Daimary faction (NDFBRD), Karbi Longri North Cachar Hills Liberation Front (KLNLF) in Assam and National Socialist Council of Nagaland – Isak Muivah faction (NSCNIM) in Nagaland. Ceasefire Agreement has been signed with the National Socialist Council of Nagaland – Khaplang faction (NSCNK) and National Socialist Council of Nagaland – Khole Kitovi faction (NSCNKK) in Nagaland and Suspension of Operation (SoO) agreement with United Progressive Front (UPF) and Kuki National Organisation (KNO) in Manipur. Also, Memorandum of Understanding (MoU) has been signed with three Meitei insurgent groups in Manipur United Revolutionary Front (URF), Kangleipak Communist Party – Lamphel (KCPL) and Kanglei Yawal Kanna Lup (KYKL) (ibid).

On February 5, 2011 the ULFA announced that it was willing to hold talks with the government of India, respecting the wishes of the people of Assam (Mathur, 2011: 269). Not only did the ULFA want to acknowledge the larger sentiment of the public, but it also had little support for the outfit's goal of a sovereign state and it was the best remaining choice which it could look forward to. The early 1990s and the 2000s were a difficult phase in this regard as the ULFA was a group which was difficult to negotiate. The outfit found safe haven in neighbouring countries like Bangladesh, Bhutan Nepal along with the tacit support of the ISI from Pakistan (SATP: 2000).

The most violent of these years of insurgent activity came in 2007-09 where serial blasts in 2008 by the National Democratic Front of Bodoland (NDFB) which targeted civilians, marked the transition from insurgency to terrorism. Such acts also showcased that rebels in the state had lost their mandate and resorted to violence against civilians in order to create fear.

As already noted that the diplomatic efforts undertaken by the Indian government, a renewed hope towards creating sustainable peace and opportunities for economic development for Assam and the north east region was now possible. Bhutan which was first unwilling to engage in armed operations to destroy approximately 30 militant camps in its territory, launched Operation All Clear in 2003 to clear out all insurgent camps following the killing of Bhutanese civilians in Bhutan by these insurgents as it posed risks to its internal security as well as India's (Mazumdar, 2005: 579). The operation along with the logistical support of the Indian army destroyed all camps killing and arresting 650 insurgents approximately (ibid: 576). Since then Bhutan has been sealed to insurgency and has been a reliable partner in countering insurgents and sharing intelligence.

Bangladesh adopted a change in its support for the insurgents (ULFA and NDFB) following the re-election of Prime Minister Sheikh Hasina Wazed who played an important role in handing over insurgent leaders such as Arabinda Rajkhowa (ULFA) in December 2009 (Hindustan Times: 2009) and Ranjan Daimary (NDFB) in May 2010 which also resulted in the seemingly better bilateral relations between India and Bangladesh (TOI: 2010). Northern Myanmar remains as a home for the National Socialist Council of Nagalim (NSCN) and some of the remaining factions of the ULFA. While the NSCN's role in Nagaland has considerably diminished, Myanmar still serves as a breeding ground for insurgents in its territory which also includes Meitei groups of Manipur. The Mutual Legal Assistance Treaty signed between India and Myanmar shows that efforts are underway and a framework has been prepared for investigation, prosecution, prevention and suppression of crime – including those relating to terrorism (TOI: 2010).

Counter insurgency operations which were intensified by the Indian Army and the Assam Police during the years 2007-09 crippled the operating capabilities of the elite category 'A' and 'C' of the ULFA which led to their surrender (Hussain: 2008). Similarly in May 2009, the Assam Police arrested Pradeep Terang, Chairman of the Karbi Longri North Cachar Hills Liberation Front (KLNLF) which led to the surrender of many of its cadres (Mathur, 2011: 275; Das 2010).

The Indo-Naga peace framework signed between the Centre and NSCN (IM) on August 3, 2015 is significant in the context of building peace in north east India. This is also significant because it has shown flexibility and realism in the approach adopted by the NSCN in terms of resolving the conflict and its willingness to alter the earlier goals of complete sovereignty and greater Nagalim to the acceptance of a constitutional framework where provisions for greater autonomy for the Naga inhabited areas were made through the establishment of autonomous district councils. The demand for the integration of greater Nagalim had been a sticking point in negotiations where Arunachal Pradesh, Assam and Manipur had categorically opposed division of territory from their respective states. It is in this context that the government of India made a proposal for a supra-state structure in 2011. This involved the granting of greater autonomy for Naga inhabited areas without territorial division of the other states. Signing the accord also meant that the support provided by the civil society to the NSCN (IM) had been persistent on a peaceful path to conflict resolution in Nagaland (Goswami: 2015).

Shifts in the policy of the government in negotiating peace in north east has been noticed since the NDFB-S killed about 70 civilians in Assam in December 2014. The government of India has now decided that it will not engage in peace talk with any outfit which involves in killing civilians and would treat such groups as terrorists. Such hard resolves from the Centre's end has resulted in the beginning of full scale operations against the NDFB-S in which senior leaders and arrests of more than 30 cadres and also some commanders were also achieved.

Civil society in north east India has also engaged in the peace building processes. Civil society is primarily an element of a society that is outside the government and other actors who cannot be regarded as an important player in peace building initiatives between the government and non-state/warring parties. It can also not be expected that the civil society is capable of managing and resolving conflicts in all situations. Civil societies often remain in the background and much of its work is limited to bringing the conflicting parties to sit together and find a negotiated solution to the problem (klingelhofer and Robinson: 2001). It should also be noted that the government of India has been insisting on holding talks directly with insurgent groups in north east India.

Civil society in north east Indian can be categorized into four groups (i) groups or bodies which claim to represent a particular ethnic community and thereby get involved in conflicts in the region, (ii) organizations which are expressly setup to make preparations for peace negotiations between two or more conflicting parties which also look to facilitate cease fire agreements (iii) groups and organizations like the church, mothers vigilante groups such as those in Manipur and Nagaland and (iv) those organizations which are loosely organized and are still in their early stages of development (Das: 2007).

Civil society organizations like the Asom Sahitya Sabha and the Asom Sattra Sanmilan have played an important role in brokering peace between the ULFA and the Government (both central and the state). In September 2004 a two day Jatiya Mahasabha was organized by the Asom Jatiyadabadi Yuba Chatra Parishad (AJYCP) which adopted two major resolutions. The first called for the initiation of talks between the ULFA and the Government while the second pressed for the ULFA to move towards reducing the armed activities of the ULFA and enter into dialogues which would be of greater benefit to the people living in the state. Such initiatives are noteworthy as these seemingly prompted banned leaders such as Paresh Baruah to agree for talks with the government in which the issue of sovereignty would form the core of the talks. The ULFA also formed the People's Consultative Group (PCG) to prepare the groundwork for eventual talks with the government in September 2005 and this resulted in the much awaited talks which were held on October 2005, the talks ended with a positive note with both the government and the ULFA agreeing to keep all channels open for further dialogues. However, the PCG withdrew from the peace talks in September 2006 citing the Government's lack of sincerity in the talks with ULFA.

Civil society initiatives in Manipur are also important where the Meira Paibis have and continue to play an important role in fighting drug-abuse, alcoholism and also staying at the forefront in agitating against human rights violations committed by the armed forces and the insurgent groups. Besides the Meira Paibis, there are other civil society groups such as the Kuki Women's Organisation, Kuki Mother's Association and the Naga Women's Movement Manipur which have tried to foster peace in the conflict ridden state. These groups including the United Committee of Manipur have led the agitation against the AFSPA since 2004. Notably Irom Sharmila also has been part of this agitation and has completed over 10 years of fasting against the draconian Act.

In Nagaland the Naga Hoho, The Naga Mother's Association and the Naga People's Movement for Human Rights have made concerted efforts in pressurizing the NSCN(IM) to settle for peace talks whilst also documenting human rights violations committed by the state and no-state armed groups. They have also applied pressure on the various Naga factions to work through a negotiated settlement for peace in the state.

Tripura has also witnessed the contributions of civil society towards the establishment of peace. Some of the notable ones in the state are Jamatiya Hoda which resolved not to pay any tax to the tribal insurgent groups operating in the state. Besides the Jamatiyas, the Reangs and the Uchai communities have joined the anti-insurgency movements. NGOs such as the Tripura Atimjati Sevak Sangh and the Borok Human Rights Group have played an important role by organizing awareness campaigns, workshops and seminars for ordinary citizens on conflict resolution which can be seen as significant efforts by the civil society in Tripura towards establishing peace.

In the context of north east India, the idea of human security is very interesting as it puts the concept of human development at the forefront of analysis in the region in present times. The region although has been granted special category status and there is the flow of funds from the centre to each of these states is perhaps a mere eyewash in the context of the region's people. Two perspectives are clearly present in this context (i) the views of the state and (ii) the views of the local people. The government of India may boast of allocation of developmental funds with a special focus but on the other hand the allegation that the region has not developed is also true. While the idea of human security is to ensure the freedom from fear and want, the region clearly lacks development and also in the context of the two fundamental aspects of human security. Life in the region is beset with fear of violence and the yearning for peace is ever present. The idea of freedom from want is something which the people of north east may have a long way to go. 'Want' in the context of north east arises from lack of development following which the problem of unemployment persists in the region. Scholars who had worked on human security issues in north east India conclude the following points which are a threat to their security and those initiatives which may result in development in this context. In this they identify that (i) armed groups are the primary sources of threat, (ii) they also conclude that the state is also a source of threat which arises through counter insurgency measures and in the process of maintaining law and order in the region. Based on these local perspectives many scholars believe that if there is strengthened emphasis on the conflict resolution process in which (i) insecurity caused by weak governance provides a fertile ground for violence, (ii) the repealing of certain provisions of the AFSPA which would result in the reduced power of the security forces and also reduce impunity, and (iii) the local population expect conflicts in the region be resolved through dialogue and not through repression. In the above points, it is evident that human security in the context of north east India, the usual locating of the regions problems are paralleled with conflicts with the resultant lack in development and infrastructural facilities. The Indian government may boast of a number of initiatives which look to improve the situation but along with this the government also carries the misconceived idea that any developmental initiative is possible in the region. Such attitudes are sought to be achieved through brute force or through repression. The ground realities and the local population's views and opinions are often ignored and governmental initiatives are sought to be implemented through force which often attracts violent protests and again the government seeks to suppress these protests through the armed forces that are enabled to act with impunity through legislative measures such as the AFSPA. The categorization of the region as north east itself presents the attitude of the Indian state in dealing with the problems which beset the eight states. The alienation of the region by such attitudes is ever present since the integration of the region with the Indian union. Thus, in conclusion it should be noted that human security in the region is essentially a significant and important aspect for building peace and also to bring development in the region. However, such initiatives should be undertaken by taking into account the local populations needs and opinions which clearly is important to prevent potential future possibilities of violent conflict in the region and ensure lasting peace.

In order to address the issue of peace building in the region, a renewed determination and adoption of different approaches are required. This is important because the solution to the problems of north east India may not lie in meeting the demands of the insurgent groups or rebel groups or with meeting the demands for altering the state boundaries. Both the central government and the state governments have the responsibility to vie for an alternative approach towards establishing sustainable peace and the recurrence of violence. Although the grim cycle of insurgency might have diminished, yet there are states like Manipur where insurgent activity still has its grip. The security establishment has been working to dismantle the strongholds in such regions and it is hoped that the region will move towards a peaceful society in the near future. However, factors like Maoism and Islamism which have seen an increasing activity in the region poses serious concerns over what lies ahead for the future of peace in the north east. Along with security measures it is imperative that the state engages in developing a strong action plan to develop the interior regions of the north east where development has not even begun. Such initiatives also facilitates the improvement of human security where the freedom from fear and want are guaranteed.

REFERENCES

Das, Biswajyoti (2010) '*KLNLF Rebels Surrender in Assam*', http://in.reuters.com/article/idINIndia-46094620100211. Accessed June 17, 2016.

Das, Samir Kumar (2007) '*Conflict and Peace in India's North East: The Role of Civil Society*', Policy Studies Brief No. 42, Washington D.C: East West Centre.

Denskus Tobias (2007) 'Peace Building Does not Build Peace', *Development in Practice*, 17(4/5): 656-662.

Doyle, Michael W and Sambanis, Nicholas (2000) 'International Peace Building: A Theoretical and Quantitative Analysis', *The American Political Science Review*, 94(4): 779-801.

Featherston, Beth (2000) '*From Conflict Resolution to Transformative Peace Building: Reflections from Croatia*', Working Paper 4, Bradford: Centre for Conflict Resolution.

Freddy, Haans J (2016) 'Engaging Youth: Challenges and Opportunities towards Building Peace in North East India', *Man and Society: A Journal of North East Studies*, ICSSR-NERC, 13(1): 106-120.

Ghosh, Atig (2013) '*Governing Conflict and Peace Building in India's North East and Bihar*', Core Policy Brief 08/2013, Oslo: Peace Research Institute Oslo.

Goodhand, Jonathan and Hulme, David (1999) 'From Wars to Complex Political Emergencies: Understanding Conflict and Peace Building in the New World Disorder', *Third World Quarterly*, 20(1): 13-26.

Goswami, Namrata (2011) '*A Non-Territorial Resolution to the Naga Conflict*' IDSA Strategic Comment, Nov, 15, 2011. Accessed, June 15, 2016.

Goswami, Namrata (2015) '*The Naga Peace Accord: Why Now?*', IDSA Strategic Comment, August 7, 2015. Accessed June 15, 2016.

Hindustan Times (2010) '*ULFA Chairman Rajkhowa Held in Bangladesh, Flown to Delhi*', Hindustan Times, December 9, 2009, http://www.hindustantimes.com/ULFA-Chairman-Rajkhowa-held-in-Bangladesh-flown-to-Delhi/Article1-482481.aspx. Accessed June 17, 2016.

Hussain, Wasbir (2008) '*The ULFA Mutiny*', Outlook India July 03, 2008, http://www.outlookindia.com/article.aspx? 237821. Accessed June 15, 2016.

Jenkins, Rob (2013) '*Peace Building: From Concept to Commission*', New York: Routledge.

Klingelhofer, Stephan and Robinson, David (2001) '*The Rule of Law, Custom and Civil Society in the South Pacific in the 21st Century: Challenges and Opportunities*', Washington D.C: International Center for Not-for-Profit Law.

Ladwig, Walter III (2009) '*Managing Separatist Insurgencies: Insights from North Eastern India*', Paper Prepared for the International Studies Association Conference, New York, 15-18 February 2009.

Llamazares, Monica (2005) '*Post War Peace Building Reviewed: A Critical Exploration of Generic Approaches to Post-War Reconstruction*', Working Paper 14, Bradford: Centre for Conflict Resolution.

Mathur, Akshay (2011) 'A Winning Strategy for India's North East', *Jindal Journal of International Affairs*, 1(1): 269-298.

Mazumdar, Arijit (2005) 'Bhutan's Military Action against Indian Insurgents', *Asian Survey*, 45(4): 566-580.

Ozerdem, Alpaslan and Lee, Sung Yong (2015) '*International Peace Building: An Introduction*', London: Routledge.

Rajagopalan, Swarna (2008) '*Peace Accords in North East India: Journey Over Milestones*', Washington D.C.: East West Centre.

Times of India (2010) '*Bangladesh Hands Over NDFB Chief, Ranjan Daimary to BSF*' Times of India May 01, 2010, http://timesofindia.indiatimes.com/india/Bangladesh-hands-over-NDFB-chief-Ranjan-Daimary-to-BSF/articleshow/5879908.cms. Accessed June 19, 2016.

Times of India (2011) '*India, Myanmar to Sign Legal Aid Agreement*', http://timesofindia.indiatimes.com/India-Myanmar-to-sign-legal-aid-agreement/articleshow/7158194.cms. Accessed June 17, 2016.

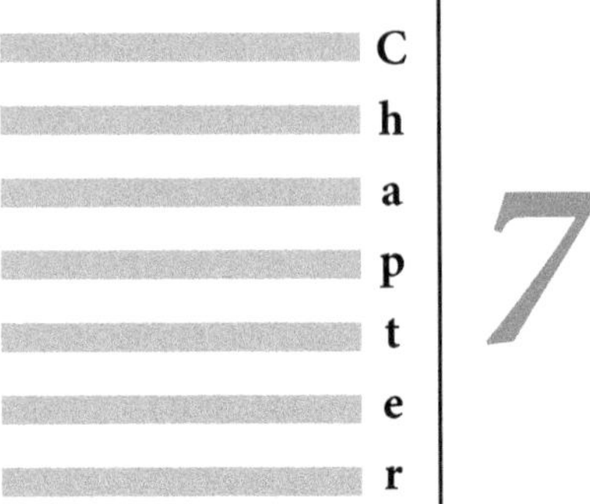

Conflict Transformation in North East India

Concepts of conflict transformation have been used to address conflicts during the post Cold-War era (Ropers 2008: 2). Generally, conflict transformation refers to the processes of transition from destructive to non-destructive conduct or how conflicts in a region have witnessed change over time (Kreisberg 2011: 50). In order to understand how conflicts transform it is pertinent to understand three basic characteristics of conflicts itself. Firstly, conflicts are generally fluid and pass through a series of stages the emergence, escalation, de-escalation and its settling down are noted. It can also be that these stages of conflict can form the basis for peace or upon which new conflicts arise. Thus, it incorporates many conflicts which are small with varying stages which include frequent backward steps. Conflict transformation includes multiple stages where the conflicting parties experience change within themselves which opens opportunities for mutually exploratory moves between them which are followed by actions that indicate mutual accommodation has begun and this is subsequently followed by building increased mutual understanding and trust. A traditional view of explaining conflicts transforming from destructive to constructive conflict is that one of the conflicting parties is responsible for the existence of such conflict and defeating that group will transform the relationship between the adversaries and constructive measures will follow (ibdi: 52-53). North east India which is host to a number of conflicts is an interesting case where conflicts in the region has witnessed varying stages of escalation, de-escalation, peace efforts undertaken to prevent violence *etc.*

Comprehending conflicts in north east India has never been easy. Conflicts in the region have lingered for over five decades since the independence of India and they have become more complex (Phanjoubam 2015: 11). Over the years conflict in the region has witnessed significant changes where shifts in policy orientation of the Indian government's policy towards the north east have been altered. In this chapter, I examine how conflict in north east India has transformed itself over the years more particularly in the last decade. Questions which are pertinent in this regard are (i) have counterinsurgency operations in the region been effective? (ii) are people in the region conflict weary and are opting for a peaceful solution? (iii) have insurgent groups lost their support base or has there been a loss of ideology among them? Further, I posit that armed insurgencies are weakening in the region and moving towards working out a peaceful solution to the problems which need to be addressed. What is important in this context is how the Indian state has gained legitimacy to impose its authority over the region. A brief description of how the region became part of the Indian Union is necessary to fully understand the dynamics of conflict in north east India. It is also important to understand how the conflict over five decades has sustained itself and how it has transformed itself.

Historically the north eastern region of India has never been part of the Indian union. However, this region was forcefully annexed by the British (Bhaumik 1998: 311). Ever since the independence of India in 1947 this region which is multi ethnic, has been vociferously demanding secession (Innoue 2005; Vadlamannati 2011: 606). In the context of the colonial background, a cursory examination demonstrates its ubiquity and its effects, even after the dismantling of the colonial empire more than five decades ago which has played out in a dramatic and tragic fashion. The colonial legacy can be identified in two lights: (i) collective historical trauma and (ii) a causal variable that continues to have an impact on the ways in which the outlook of states has been, post decolonization (Miller 2013: 8). Between the years 1874 and 1934, colonial policies segregated the region where tribal populations were administered under the category of non-regulated, backward or excluded areas. Such categorization of the region prevented all outsiders to enter these areas except those who obtained special permission from the government under the Inner Line Regulation of 1873. While an extension of this regulation to almost all hill areas created a frontier within a frontier which highlighted the political and cultural rift between the people of the plains and the tribals living in the hill areas. These factors created a situation where the tribal areas were excluded from the administrative patterns which existed in other regions of the country. The Government of India Act of 1935 also continued with this policy of exclusion due to which the overriding effects were twofold. The integration of the many tribes and communities which could have been facilitated by the British was lost on the one hand while on the other, tribal communities continued to be excluded from the socio-political developments which were taking place elsewhere in the country. Thus, such exclusionary policies which existed during the colonial rule, and continued to exist during the post independent India, created conditions for ethnic conflicts due to unequal and unbalanced development of the region. The response of the Indian state to these problems was the creation of autonomous districts and regions which were often identified with tribal affiliations through constitutional measures such as the sixth schedule. Many of these regions subsequently became full states which resulted in the demands for such similar arrangements for homelands to be made by many of the tribes living in the region (Upadhyay 2006: 4993).

In the words of Ashild Kolas, the pervasiveness of conflict in north east India is often attributed to the regions underdevelopment and its natural tendency to engage in violence among its indigenous population. Employing a standard frame researchers and local analysts on north east India often describe the region as a site of ethnic conflict (Kolas 2015). Challenging such perspectives scholars like Sajjad Hassan argues that conflicts in the region are fuelled by malgovernance, failure to provide security to the people, ensure transparency and accountability and also address economic disparities. Although ethnicity is often a mobilizing factor conflicts in north east are not about the inherent differences between groups but the absence of an effective medium to regulate these relationships (Hassan 2007; Hassan 2008).

As it is evident, among the eight states of north east, Assam, Manipur, Nagaland and Tripura are those that have shown greatest propensity towards violent conflicts relating to secessionist movements and ethnic conflicts. Although violence seems to be waning weak, Manipur is one state which is experiencing high levels of violence. The state still continues to have violence which involves bloody clashes among ethnic groups whilst often experiencing breakdown of governmental authority. On the contrary Mizoram is one example where violence has been virtually absent for over three decades and inter-group contestations seem to have been better managed. Although, conflicts in Mizoram also revolved around ethnic identity and nationalism as in other states in north east India, it is interesting to note that the outcomes of such movements in north east has not been similar vis-à-vis Mizoram. The restoration of peace in Mizoram can be attributed to the readiness of both the central government of India to accommodate the demands of the Mizo National Front (MNF) and its economic largesse towards the socio-economic development of the state (Baruah 2005: 71). What is interesting to note here is that such arrangements have not been able to achieve desired goals in other states where armed conflicts were present.

While the Khalistan movement in Punjab abated by the end of 1992, the north eastern region witnessed significant increase in violent incidents which were related to the creation of autonomous districts councils, redrawing the boundaries of states to create new states, and also demands for autonomy. These conflicts included various ethnic groups clashed among themselves – Kuki-Naga Conflict (1993-98), tensions between the Bodos and Adivasis (1996), Dimasa-Hmar (2003), Kuki-Karbi (2003), Dimasa-Karbi (2004), Rabha-Garo (2011) and the Bodo Muslims (2010-2011) were active (Haokip 2013: 78). The formation of militant groups with secessionist demands such as the National Socialist Council of Nagaland-IM (NSCN-IM) and the United Liberation Front of Assam (ULFA) in 1980 and 1979 marked a shift in the way demands in north east India were being pursued. Although it might be noticed that there are many differences, militant groups in north east India have all been formed towards achieving sovereign nationhood, independence or autonomous homelands. While demands of territorial claims overlap, inter-group violence is often related to disputes which relate to economic stakes in taxation difficulties in maintaining alliances, rivalry between armed groups and factional fighting (Ashild 2015).

One of the primary objectives of counterinsurgency operations is to restore peace in the region in which it is conducted. It is also important to recognize the fact that every conflict which is internal requires a different notion of victory than conventional victory (which is rather rare) by the government. In some cases, conflicts just abate and rebels just revert to the usual normal life in society while in some other cases negotiations between conflicting parties reach a political solution accommodating interests which are mutually fulfilled. Importantly, success will imply that conflict has been reduced to a manageable level of threat where the government looks at qualified success which is meant as the end of political threat.

In democracies such as that of India, it is imperative that to restore peace, the state ought to achieve some sort of reconciliation with those rebels and successfully integrate them back into the society. In the context of north east India, the government often viewed the rebels as being misled and often confused rather than enemies to the state. However, this is not to refute the underlying grievances by which the rebels received motivation to engage in armed violence against the state and the occasional failures of the state forces to exercise restraint in using force, but it however underlines the fact that the government of India wanted to see the rebels reconciled with the state. This showcases the attitude of the Indian government which has been attempting to minimize the use of force and maximize the use of political compromise (Ladwig 2009).

The years 2004-05 seemed to signal both hopes for a reduction in conflict in north east as well as its escalation between various ethnic groups in the region. This is often attributed to the regions underdevelopment and its peripheral positioning with the Indian geography. The inauguration of the 789 car rally in 2004 and the tentative peace initiatives which were initiated seemed to again signal major changes in the region's conflict situation (Barbora 2006: 3805). In the north east where two dominant groups which grab the attention of any casual observer or those interested in the conflict in the region, are the NSCN-IM and the ULFA. However, it is also important to note that the NSCN-K (Kaphlang) in recent years has become a force which cannot be ignored in the context of the conflict in Nagaland. Despite the complications in terms of dealing with two factions, the government of Indian has still managed to sign ceasefire agreements. The NSCN-IM agreed to ceasefire in the year 1997 and the NSCN-K in 2000 which have been unilateral and neither has been particularly effective. However, peace talks between these groups have been taking place seriously since then. One of the most potentially promising developments through these measures is that the NSCN-IM has overtly declared withdrawing the demand for a sovereign nation for the Nagas, although the demand for greater Nagaland continues to be one of its objectives (Cline 2006: 133). However, it should also be noted that the NSCN-IM's leader Muivah has stated that the outfit has not given up its demand for sovereignty (The Hindu 2016). He also claimed that the concept of shared sovereignty has been achieved as the unique history of the Nagas is recognized. He however, maintained that when this would be implemented is to take time and it has not been exactly determined (ibid).

The resolution to the long Naga problem was seen to be nearing with the recent signing of the Framework Agreement between the Indian government and the NSCN-IM on 3 August 2015. Although the contents of the framework still are not known, it is expected that this initiative will bring change in the region. In this regard the NSCN-IM has been capable of galvanizing support among other political and social entities such as the Naga Hoho, The Naga Students Federation, The Naga Mothers Association, The Naga Peoples Movements for Human Rights and the United Naga Council, for a negotiated settlement with the government of India (Chhonkar 2016). The NSCN-IM has also been trying to re-establish it dominance in the Naga inhabited areas in the states of Arunachal Pradesh, Assam and Manipur. In this context, the proposition to develop a supra-state or non-territorial unification of the naga inhabited areas in the states of Manipur, Assam and Arunachal Pradesh seems to be a meaningful way of overcoming the ethnic tensions. However, in contrast this has been a highly emotive issue where the territorial unification of the Naga inhabited areas has been rejected by these states (Goswami 2011). It has also been upping its efforts to reshape the existing the loyalties in eastern Nagaland following the defection of General Khole Konyak from the NSCN-Khole Kivitovi (NSCN-KK) (ibid). Tracing the splits in the NSCN-IM, it is interesting to note that with every round of negotiations with the government of India, atleast one faction has emerged resulting in the limiting of the sphere of dominance by the NSCN-IM. This has also resulted in faction fighting between the three groups (IM, Kaphlang and Khole-Kivitovi) which has again resulted in smaller faction being formed. The NNC has also been divided into smaller factions. These splits in the Naga Resistance movements point at two directions where the NSCN-IM has been able to dominate negotiations with the Indian state in terms of the Manipur state's hill districts and on the other hand the Government of India has not been able to decide with whom it has to negotiate to establish sustainable peace as all Naga factions have their own support base and agenda which differ with each other – hence a complex puzzle which cannot be easily comprehended. In this scenario the Indian government has to take into consideration each faction's demands and make a proposal which would be inclusive of

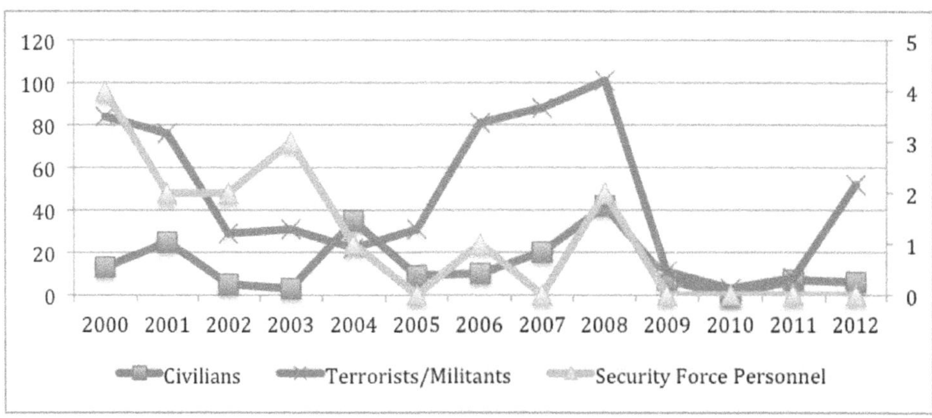

Fatalities in Nagaland, 2000-2012.

Source: Vasundhara Sirinate.

all the groups' demands - which is clearly an impossible task. This absence of clarity has in some sort enable the state of sign ceasefire agreements with the NSCN-IM and it has also been encouraging other dominant groups to enter such agreements (Sirinate 2015).

The figure above clearly indicates that there has only been a reduction of casualties in among the security forces due to the signing of the ceasefire agreements. Civilian casualties were at the highest during the year 2008 after the ceasefire agreement was extended. Looking at the casualties among the insurgent groups it is clear that these fatalities were not due to the operations conducted by security forces as it indicates that there was cessation of hostilities and the ceasefire was in place. These fatalities can be explained by looking at the fighting which existed during this period between the various factions of the NSCN. Reports also indicated that the NSCN-IM in order to establish its dominance had often engaged in killing member of the NSCN-K. Frequent fighting between the NSCN-K and NSCN-KK also contributed to the deaths of members of these insurgent groups. The year 2009 and 2010 witnessed a drop in the deaths of insurgents due to the Covenant of Reconciliation signed between the NSCN-K, NSCN-IM and the NNC in June 2009. However, the year 2011 again saw the rise in killings as rivalries over leadership, which were internecine in nature arose. An important development at this juncture is that the ceasefire agreements put the NSCN-IM at the forefront of negotiations and also put the burden on the group to convince other groups to the negotiations (Sirinate 2015). The events following are evidence to the fact that the Government of India and the NSCN-IM seem to have agreed upon a framework for peace in order to resolve the conflict amicably.

The state of Assam seems to be experiencing relative peace following decades of armed conflicts and violence. Since 2010, there has been a consistent decline in terms of insurgent violence. The state has also witnessed a sudden reduction in terms of fatalities which were recorded at 437, 370 and 391 in the years 2007, 2008 and 2009. By the year 2010 civilian casualties had also reduced considerably with 664 over the years 2007, 2008 and 2009 with an average of 221 per year to just 48 killings during the year 2010. The transformation of conflict in Assam to what it is now is reflective of the fact that peace can be established rather quickly with the cooperation of neighboring countries who have provided safe havens within their country. This is a point which has come to reassert itself in the case of Assam where the Bhutan Royal Army almost decimated the National Democratic Front of Assam (NDFB) in 2003. With the decision of the Bangladeshi Awami League Government to turn over several top leaders of the United Liberation Front of Asom (ULFA) has forced the outfit to engage in negotiations. However, until the handing over of the ULFA's Chairman, Arabinda Rajkhowa, in December 2009, the prospects of negotiations for peace remained bleak (Routray 2011). Another significant development in this regard was the handing over of the anti-talks faction (NDFB) leader, Ranjan Daimary to the Border Security Force by Bangladesh in 2010 (TOI 2010). Ranjan Daimary was the prime accused in the serial blasts which occurred in Assam in 2008 (ibid).

The year 2009 proved to be of significance for the state of Assam in term of the conflict. Amongst the armed groups in Assam, the Dima Halam Daogo (DHD) remained the most elusive and violent groups which was operating in the north Cachar hills and Karbi Anglong Districts. The DHD which was formed following the surrender of the Dimasa National Security Force (DNSF) with the exception of its commander Jewel Garlosa who did not surrender - formed the DHD. The DHD demanded the formation of the Dimaraji which comprised of the Dimasas who inhabited the North Cachar Hills, Karbi Anglong Districts parts of Nagaon in Assam and Dimapur and those areas occupied by the Dimasas in Dhansiripar in Nagaland. It was Jewel Garlosa who formed the Black Widow of the DHD (G) whereas the DHD was led by Dilip Nunsia who remains a rational actor and has signed a ceasefire agreement with the Indian government since 2003 which is still in force. The leader of the DHD (G) Jewel Garlosa was arrested in Bangalore in the year 2009, following which the 386 heavily armed youth of the faction surrendered where the Commander-in-Chief declared in the surrender ceremony on October 2, 2009 that the North Cachar Hills will remain peaceful from this day forward and the east west railway corridor will be completed with its cooperation. This is a significant factor in terms of the development of the remote areas of Assam. The DHD had until its surrender had been disrupting construction work and killed workers and kidnapped those engineers who were engaged in these projects (Goswami 2009). The one most significant caveat in terms of peace in Assam seems to be the arrest of Paresh Barua who is allegedly hiding along the Myanmar - China border and operating the ULFA. However, the ULFA which has lost its popular support and political propaganda machine, it is difficult to imagine if Paresh Barua can single handedly manage and revive the ULFA. Perhaps this marks the end of the ULFA (ibid).

Among all the states of north east India, Manipur continues to have significant unrest. During the year 2005 Manipur witnessed an upsurge in violence. The state is host to approximately 30 separate insurgent groups which has about 10000 members within them. It is interesting to note that insurgent activity in Manipur has at most times targeted the Capital rather than the rural areas. The level of extortion by various insurgent groups is extraordinarily high vis-à-vis other north eastern states (Cline 2006: 136-37). In 2009 estimates suggest that deaths related to insurgent violence had reduced to 369 in 2009 from 468 in 2008. Significantly 19 organizations of which 11 under the Kuki National Organization and 8 under the United Peoples Front agreed to the Mutual Suspension of Operations (SoO) after a review of the peace process under the Agreed Ground Rules. However, this did not set the stage for dialogues with the government of India and these groups as they had their own differences lingering within them. Conflicts in Manipur have remained predominantly among those insurgent groups which claim to represent their ethnic communities. The United National Liberation Front (UNLF) the groups which is Meitei centered has engaged in attacking the Thangkhuls in Ukhrul district of Manipur. These attacks were carried out from their camps in Mintha near the Ukhrul-Chandel border in Manipur (Goswami 2009).

In Manipur three dominant groups seem to be conflicting with each other. On the one hand the NSCN-IM's demand for the integration of Naga inhabited areas remains a point of contention which is opposed both by the Meiteis and the Kukis, while on the other the Kukis also demand the creation of Kuki land carving out areas from the hill of Churchandpur, Chandel, Senapati, Tamenglong and Ukhrul. Thus, the bone of contention between the Kuki's and the Naga's remains the land dispute. The demand for Kukiland remains a direct challenge to the demand for the integration of the Naga inhabited areas into Nagaland where the NSCN-IM has also claimed those districts which are demanded by the Kukis (Tungpo 2016; Piang 2015). While the Meiteis have opposed the creation of Kukiland or greater Nagaland, the Kukis or the Nagas have however not been able to coordinate with each other. Many explain the inability of to coordinate as a result of the tensions which have remained in the aftermath of the clashes between 1992-1997. While these tensions linger between the two groups the government of India engages with the NSCN-IM in political dialogue whilst ignoring calls by the Kukis for political dialogue (Kipgen 2013). In this context it might be well to reiterate the five resolutions passed unanimously by the Manipur Legislative Assembly in a span of five years (1997-2002) where the territorial integrity would not be compromised at any cost. This was brought out as a government document entitled Manipur Culture Policy 2020 which reiterated its stand on the unity and integrity of the state. This legislation triggered one of the worst bouts of violence in Manipur's recent history where 18 people were reportedly killed and many governmental establishments were set on fire. The government of India as a result of the severity of agitations withdrew the declaration and the protests subsided (Das 2010: 124).

In 2015, the government of Manipur passed three bills (i) Protection of Manipur People's Bill 2015, (ii) Manipur Land Revenue and Land Reforms (Seventh Amendment) Bill, 2015 and (iii) Manipur Shop and Establishment (Second Amendment) Bill, 2015. These bills were vociferously opposed by three major tribal student groups namely the All Naga Students Association of Manipur, Kuki Students Association and the All Tribal Students Union Manipur. The Land Reforms Bill as perceived by the Kukis and Nagas is a covert attempt by the Meiteis to gain access to the scheduled hill districts. They argue that outsiders were never a threat since they can neither own or buy or pose any form of threat to government jobs in the valley. According to section 158 of the Land Reforms Bill and Manipur Land Revenue Act of 1960, those lands which belong to tribals in the valley cannot be sold to non-tribals in the valley areas unless prior consent is granted by the concerned deputy commissioner. However, with the present Protection of Manipur Peoples Bill, Clause 2(b) defines Manipur People as persons of Manipur whose names are registered in the National Register of Citizens 1951, Census Report 1951 and the Village Directory of 1951, which has raised serious apprehensions among the tribal people in Manipur. If the bill is made law, a person needs to possess documents of registration in all three registers failing which they would be considered as outsiders – thus meaning that many of the hill people in the state would be excluded. This bill is interestingly similar to the Ceylon Citizenship Act of 1948 which placed many of the Hill Tamils in Sri Lanka as stateless people. The three Bills which were enacted in the Legislative Assembly of Manipur in 2015 seems to be much similar with a slightly different tinge where the tribal people living in the hills would suffer structural violence perpetrated against them henceforth. Hence the Sixth Schedule, at this juncture seems to offer an amicable solution where the suggested political arrangement would facilitate the autonomy for the Kukis and the Nagas without territorial division of the state of

Manipur (Kipgen 2015). Thus until recently the state of Manipur has remained with active conflicts which have been highly destructive where peace and security for the civilian has been disturbed to a large extent.

Bounded by more international borders, Tripura shares its border in the north western, south western, western and northern boundaries with Bangladesh. Unlike other states in north east, the state of Tripura was integrated with the Indian Union peacefully when it acceded to India on 13 August 1947 and subsequently chose to integrate with India fully in the year 1949. However, Tripura is another state which illustrates the stark realities of ethnic colonization with the Indian Union. Revolts in Tripura have been due to the fear of the loss of identity because of such encroachments. Tripura did not experience insurgency during the time of independence. It much like Mizoram began its Journey with the Indian state on the strong foundation of legal and moral accession. However, grievances which arose after years of neglect and misrule and an attitude of partisan governance by the ruling elite caused dramatic changes in the demographic composition of the state where the tribal population reduced to one-third from two-thirds of the total population. The seeds of insurgency were thus sown and the first act of frustration and dissatisfaction arose from the Tripura Upajati Juba Samatiya (TUJS) in June 1967. Their demand was the creation of autonomous district councils under the provision of the sixth schedule of the Constitution, recognition of Kok Borok as the official language and restoration of tribal lands that had been alienated followed by the development of the region. Subsequent years witnessed the rise of Bijoy Harangkhawl who headed the TUJS. After the TUJS failed to achieve any of its stated objectives the Tripura National Volunteers (TNV) was formed with the aim to fight for Tripura's freedom on 10 November 1978. The TNV operated from its bases in the Chittagong Hill Tracts, whilst the TUJS continued to fight for their demands since they were not satisfied with the recognition given under the Seventh Schedule of the Constitution. However, a compromise was arrived at after the formation of autonomous district councils which were specified in the Sixth Schedule of the Constitution in July 1985. More than a decade of violence, Harangkhawl did not accept peace even after the Memorandum of Understanding with the Government of India was signed on 12 August 1988. Following these developments the state also witnessed the emergence of two more insurgent groups (i) National Liberation Front of Tripura (NLFT) and (ii) All Tripura Tigers Front (ATTF), in 1989 and 1990. These two groups were involved in bloody and protracted conflicts with the government and occasionally there were also surrenders from the two groups. Despite these episodes of violence in Tripura, there has been a reduction of armed conflict in the state and it has witnessed peace to a certain extent although there were sporadic events of violence which still continue to exist (Chadda 2005: 354-56). Between the years 1990 and 1995, insurgency in Tripura remained low however, the years 1996 to 2004, it grew in magnitude after which it slowly thawed out. It is important to examine what made these movements to slowly wane away. The sagacious and visionary leadership of Chief Minister Manik Sarkar formulated a multi-dimensional strategy to constructively respond to the insurgent situation in Tripura. In contrast to other states in north east India, counter insurgency operations did not involve the Indian army but the para-military from the center and the state police force were forged to form a synergetic, coordinated and cohesive mode to derive maximum gain. This helped in very minimal violation of human rights which fighting insurgents and which helped in reducing animosity cropping up in the minds of the local people against the operations or the security forces. The political initiatives undertaken by the Chief Minister Manik Sarkar through peace marches enabled the state to instil confidence in the minds of the people that the state was sincerely committed to developing all segments of the state. Further, these initiatives included strengthening, revitalizing and legitimizing micro-development organizations such as autonomous development councils, gram-panchayats and village councils. Such initiatives brought significant empowerment particularly among the tribals in the state. Thus the state has been able to script its victory and proved that insurgency is not an insurmountable phenomenon (Sahaya 2011; Garg 2016). One of the most significant moments in the history of Tripura is the repeal of the Armed Forces Special Forces Act. The repeal is evidence to prove that insurgency and armed violence is waning weak in the state (Ali 2015). However, it may be too simplistic to draw such conclusions as every insurgency has its own complexities which may not accept such initiatives as solutions to their demands.

Although it is debatable, it can be posited that conflicts in the region is weakening by arguing that insurgent groups are fast losing its support base and seem to be moving towards a peaceful solution to conflicts in the region. More particularly the recent Naga Peace Framework which was signed between the NSCN-IM and the Government of India seem to be pointing towards this direction. It should also be noted that scholars in recent years are looking at conflicts in north east India with an alternative approach where youth participation and their motivations to engage in conflicts are taken into consideration (Freddy 2016). Weakening of conflicts in the region is also due to the fragmentation or breakup of the dominant groups into smaller factions. This also is indicative that existing disagreements within these groups lead to the creation of smaller factions which at most times fail to make a significant impact towards their stated objectives. Furthermore, the recent death of Isak Swu of the NSCN-IM is also a fact pointing towards the slowly diminishing dominance of the group in the region. This factor in the history of the NSCN-IM seems to be pointing at the future leadership of the group. At present the NSCN-IM is headed by Muivah and there seems to be no indication that a probable successor to the group has been looked into. This again leads to the conclusion that there may be more breakaway factions from the NSAN-IM thus eventually leading to the conflict abating.

In the context of Manipur, the conflict situation seems to be similar to that of Sir Lanka where Sinhala Chauvinistic tendencies may be compared to the Meiteis. Meitei Chauvinistic tendencies may be found in the recent Bills on (i) Protection of Manipur People's Bill 2015, (ii) Manipur Land Revenue and Land Reforms (Seventh Amendment) Bill, 2015 and (iii) Manipur Shop and Establishment (Second Amendment) Bill, 2015 which seem to favor Meitei dominance in the state which also sidelines the tribals living in the hill of Manipur resulting in structural violence being perpetrated against minorities. Thus Manipur seems to be continuing with various forms of violence and conflicts during the past decade. The nature of newer conflicts also seems to be signalling a combination of structural and direct violence in the region while a reduction of secessionist movements may be observed.

Tripura presents a classic example that insurgencies can be managed through inclusive development initiatives which would result in conflicts being overcome and a return to peace. The withdrawal of AFSPA can also be seen as a significant development in the politics of the state which might also be the first step towards the repealing of the Act in other state of north east where conflict and armed violence is slowly but substantially reduce. Thus in conclusion, conflicts in north east India are waning weak and there may be a return to peace in the near future paving way for substantial development of the region.

REFERENCES

Ali, Syed Sajjad 2015. Tripura Withdraws AFSPA, Says Insurgency on the Wane', The Hindu 28 May 2015.

Barbora, Sanjay 2008. 'Rethinking India's Counter-Insurgency Campaign in North East', *Economic and Political Weekly*, 41(35): 3805-3812.

Baruah, Sanjib 2005. *'Durable Disorder: Understanding the Politics of North East India'*. New Delhi: Oxford University Press.

Bhaumik, Subir 1998. 'North East India: Evolution of Post-Colonial Region'. In *Wages of Freedom*, edited by Partha Chatterjee'. New Delhi: Oxford University Press. pp. 310-311.

Bhaumik, Subir 2009. '*Troubled Periphery: Crisis of India's North East*', New Delhi: Sage Publications.

Chadda, Vivek 2005. '*Low Intensity Conflicts in India: An Analysis*'. New Delhi: Sage Publications.

Chhonkar, Pradip Singh 2016. Naga "Framework Agreement" and its Aftermath', IDSA Comment', New Delhi: IDSA, September 1, 2016. Accessed: 25.10.2016

Cline, Lawrence E. 2006. 'The Insurgency Environment in North East India', *Small Wars and Insurgencies*, 17(2): 126-147.

Das, Samir Kumar 2010. *India: Democracy Nation and the Spirals of Insecurity: State Response to Ethnic Separatism in India's North East*, in Robert G. Wirsing and Ehsan Ahrari, (ed) 'Fixing Fractured Nations: The Challenge of Ethnic Separatism in the Asia Pacific'. New York: Palgrave Macmillan. pp. 116-139.

Freddy, Haans J. 2016. 'Engaging Youth: Challenges and Opportunities towards Building Peace in North East India', *Man and Society: A Journal of North East Studies*, 13(1): 106-120.

Garg, Ibu Sanjeeb 2016. 'Unrest in Tripura-The Struggle and the Way Forward'. The Shillong Times. 6 September 2016.

Goswami, Namrata 2009. '*An Assessment of Insurgencies in Assam, Manipur and Nagaland in 2009*'. IDSA Issue Brief. New Delhi: Institute of Defense and Strategic Analysis.

Goswami, Namrata 2011. '*A Non-Territorial Resolution to the Naga Ethnic Conflict*', IDSA Comment, New Delhi: Institute of Defense and Strategic Analysis.

Haokip, Thangkholal 2013. The Kuki-Naga Conflict in the Light of Recent Publications', *South Asia Research*, 33(1): 77-87.

Hassan, Sajjad 2008. 'Understanding the Breakdown of North East: Identity Wars or Crises of Legitimacy', *Journal of South Asian Development*, 3(1): 53-86.

Hassan, Sajjad M. 2007. '*Understanding the Breakdown of North East India: Explanations in State-Society Relations*', London School of Economics, Working Paper SeriesNo. 07-83. London: London School of Economics.

Inoue, Kyoko 2005. '*Integration of North East: The State Formation Process. In Sub-Regional Relations in Eastern South Asia: With Special Focus on India's North Eastern Region*'. Study No.133. Tokyo: Institute of Developing Economies.

Kipgen, Nehginpao 2013. '*Intricacies of Kuki and Naga Ethnocentricism in Manipur*', www.huffingtonpost.com/nehginpao-kipgen/intricacies-of-kuki-and-naga_b_2531115.html. Accessed: 25.10.2016.

Kipgen, Nehginpao 2015. '*Manipur on the Brink*', The Hindu, September 2, 2015. Accessed: 25.10.2016.

Kolas, Ashlid 2015. 'Framing the Tribal: Ethnic Violence in North East India', *Asian Ethnicity*, http://dx.doi.org/10.1080/14361369.2015.1062050. Accessed: 17.10.2016.

Kreiberg, Louis 2011. 'The State of the Art in Conflict Transformation', in Beatrix Austin, Martin Fischer, Hans J. Giessman (ed) *Advanced Conflict Transformation: The Berghof Handbook II*. Berlin: Berghof Foundation. pp. 49-69.

Ladwig, Walter III (2009) '*Managing Separatist Insurgencies: Insights from North Eastern India*', Paper Prepared for the International Studies Association Conference, New York, 15-18 February 2009.

Miller, Manjari 2013. '*Studies in Asian Security: Wronged by Empire: Post-Imperial Ideology and Foreign Policy in India and China*,' California: Stanford University Press.

Phanjoubam, Pradip 2016. '*The North East Question: Conflicts and Frontiers*'. New York: Routledge.

Piang, Lam Khan L. 2015. 'Overlapping Territorial Claims and Ethnic Conflict in Manipur', *South Asia Research* 35(2): 158-176.

Ropers, Norbert 2008. '*Systemic Conflict Transformation: Reflections on the Conflict and Peace Process in Sri Lanka*'. Berlin: Berghoff Foundation.

Routray, Bibhu Prasad 2011. '*Conflict Transformation in Assam: Lesson and Challenges*', www.claws.in/621/conflict-transformation-in-assam-lessons-and-challenges-bibhu-prasad-routray.html. Accessed: 25.10.2016.

Sahaya D.N. 2011. 'How Tripura Overcame Insurgency', The Hindu. 19 September 2011.

Sirinate, Vasundhra 2015. 'Can an Accord End an Insurgency?', *The Arena*, The Hindu Centre for Politics and Public Policy, August 7, 2015. www.thehinducentre.com/the-arena/current-issue/article7511878. Accessed: 25.10.2016.

The Hindu 2016. '*NSCN has not Given up on Sovereignty, Says Muivah*', The Hindu, New Delhi, 8 July 2016. Accessed: 25.10.2016.

Times of India 2010. '*Bangladesh Hands Over NDFB Chief Ranjan Daimary to BSF*'. Times of India, May 1, 2010.

Tungpo, Siamchinthang 2016. '*Manipur is Heading for Ethnic Conflict*', Imphal Times June 5, 2016. Accessed: 25.10.2016.

Upadhyay, Archana 2006. 'Terrorism in the North East: Linkages and Implications', *Economic and Political Weekly*, 41(48): 4993-4999.

Vadlamannti, Krishna Chaitanya 2011. 'Why Indian Men Rebel? Explaining Armed Rebellion in the North Eastern States of India 1970-2007', *Journal of Peace Research*, 48(5): 605-619.

Engaging Youth in Peace Building in North East India: Some Observations

The circumstances in which young people grow up are varied and their needs and perceptions differ. While the idea of youth does not exist as a single group, there is however, an urgent need to understand what is happening in the lives of young people. According to a young person's cultural and national context, young people have to negotiate their future and understanding the realities of youth have questions that relate to the approach, framework or perspectives that are taken into consideration (Wyn and White, 1997). The idea of youth in everyday life is often taken to mean the state of being young and that which particularly denotes to the phase of life between childhood and adulthood. The term youth often refers to ideas of being unruly, an idea that is often associated to males and at most times they have negative descriptions (Spence, 2005). Youth are important assets and they represent and reflect our hopes and aspirations and dreams for the future of our society. According to Silbreisen and Lerner (2005), young people have a significant role in accelerating changes in demography, technology, economy and politics. However, most research on youth more particularly young men, suggests that a large proportion of male youth increase the likelihood of conflict. This perspective of young men as destabilizing agents is contradictory to the fact that young people have become leaders in peace building and have contributed to the reconstruction process, despite the fact that participation of youth is higher. Young people's experiences during conflict have a profound effect on their understanding of peace and conflict and as leaders of the next generation, they therefore, have the potential to guide future processes towards reconciliation or renewed conflict. What is positive in today's world is that there is and increased appreciation in terms of young people's participation in peace building processes and is receiving more attention among policy makers and scholars (Fearon and Laitin, 2003).

It is in this context that the north east region which has been facing insurgent movements and armed conflicts for over five decades, has brought the question of building peace in the region. Various measures have been taken to address the issues that are present in the region. However, one question that remains and continues to be elusive is the engagement of youth in the peace building process. North east of India as it is popularly known, is comprised of eight states, namely, Assam, Arunachal Pradesh, Manipur, Mizoram, Nagaland, Sikkim and Tripura. Sikkim became a protectorate of the Indian Union in the year 1947 and was made into a full sate in the year 1975. The region is rich with natural resources, covered with dense forests and it receives the highest rainfall in the country. With large and small rivers flowing through the land this region is home to a cache of different fauna and flora. The region is also diverse in cultures, customs, languages and traditions which is home to multifarious social and ethnic groups. The people of north east India have high self-esteem which has made them averse to accepting the diktat of the 'imagined others' which stems from ideas of democracy and the practice of discussion prevailing among the various tribes that are present in the region. Despite such descriptions of north east India which paints a picture perfect, this region has been experiencing ethnic conflicts, low productivity and market access, poor governance and lack of infrastructure. The Government's inability to address problems caused by remoteness, seclusion, backwardness, have all provided a fertile ground for breeding armed insurgencies and various other forms of conflicts in the region that are related to identity and ethnicity. What makes the region distinct from the rest of India are the assertions of various ethnic identities and attitude of the state in containing them.

Commonly studied as a theatre of insurgency and counter insurgency by many scholars, north east India is marked by a high degree of alienation in terms of the relationship between the population of north east India and Indians on the other side of the narrow strip of land that connects the region to the rest of the country. Integration without consent, colonial attitudes, nativism, legal and illegal migration, relative deprivation, cultural nationalism and irredentism have all sparked violent conflicts that has manages to stay active in the region for more than five decades.

We have seen in the previous chapters, how youth in developing countries are engulfed by conflict or post conflict situations and how it significantly alters their lives and future prospects. Time and time again the frequency in which students or youth have been at the forefront of opposition movements and it is these events that have captured the attention of scholars and researchers globally. Around the world the mobilization of youth has been prominent across movements that are directed towards political reform. Yet there is significant variation that is seen in terms of motivations, shape and impacts of such engagements. There is however, little theoretical or comparative research that could suggest the impact of such activism although there is continued manifestation of youth participation in conflicts or protests (Weiss, Aspinall and Thompson, 2012). The process of globalization has meant that there is greater awareness of human rights abuses, degradation of the environment, political social and economic injustices and it is through such consciousness that youth activism and movements have emerged as a response to such grievances that they face within their societies (Koeffel, 2003).

In the context of north east India, there has been significant youth movements in all eight states of the region. North east India has been a hot-bed for student movements that have been on occasions at a large scale (Baruah, 2002). Youth or student movements are categorized by ethnic mobilization, cultural autonomy and exclusive possession of local resource base. Youth in north east India have passed through different ideological climates such as liberalism, conservatism, nationalism and alongside these differences they have also experienced a threat perception in terms of demographic imbalance, relative deprivation and loss of cultural identity (Chakrabarti, 2008).

What is interesting, important and needs to be understood is that youth organizations such as the North East Students Organization, Assam Students Union, Manipur Students Federation, All Manipur Students Union, Kuki Students Association, Democratic Students Alliance of Manipur, Kangleipak Students Association, Naga Students Federation, Naga Peoples Movement for Human Rights, Naga Students Union and many others are engaged directly or indirectly in conflict. These youth organizations have the capacity to cripple daily life in the region by calling for strikes and also through direct engagement in violence. Interestingly many of the militant organizations in the north east have been founded by individuals belonging to the age group of youth who had initially expressed their grievances against the Indian state. What is more fascinating is the fact that these groups enjoy local support from the local people living in the region.

Engaging Youth in Peace Building in North East India

In conflict settings, engaging youth in the peace building process in of utmost importance. When youth are engaged in the peace building process, there is the advantage that these individuals can provide new and clear insights into those complexities present in the conflict that might not have been addressed thus far. More importantly youth can be used to gather information on the actual causes for contention between conflicting parties which in turn would help policy makers to design effective strategies that would prevent a relapse into conflict. Young people who had engaged actively in conflict, for example, could provide critical information in terms of demining the region where conflict had been present. It can thus be posited that youth engagement in peace building processes can prevent a relapse in conflict.

In order to answer the ensuing question how can youth be positively engaged in the peace building process, it is necessary to characterize youth as (i) those who had actively participated in conflict and (ii) those who had either been victims or those who had not engaged actively in the conflict. While there are difficulties in terms of identifying/making the distinction between active and no-active participants in conflict there can be much achieved through such a distinction. It should also not be ignored that making such a distinction brings with it new problems where active participants in conflict maybe treated differently or they may not be accepted into the mainstream life within a society while it may be otherwise for non-active participants in the conflict.

Despite these challenges, this distinction is useful as those who have been active in conflict through indoctrination and have been used to a culture of violence may need more attention placed upon them in terms of rehabilitations and trained in skills that would be required for the return to normal life within a society.

In order to fully utilize the potential of youth while engaging them in the peace building process in north east India, the following points need to be taken into consideration: (i) training and capacity building programmes should be organized on issues of relevance to post conflict societies, (ii) governments need to ensure that youth are fully engaged in the decision making process so that the aspirations and needs of youth in such settings are fully addressed and implemented, (iii) young people should be sensitized on secularism an plurality, (iv) young people need to be sensitized on the impact and consequences of war and terrorism – violence and conflict, (v) young people need to be involved in achieving national targets and Millennium Development Goals (MDGs), (vi) initiate dialogues on democracy, peace and human rights standards between youth and adults, youth and educational institutions, (vii) promote equal treatment of women, (viii) train local youth leaders in democracy, rights of citizens an rule of law, (ix) creating job opportunities and thereby enabling youth to live with dignity and self-sufficiency, (x) encourage youth cooperation at the national level and international level, (xi) encourage youth work in reconstruction programmes in post conflict societies and (xii) provide and effective grievance redressal system and (xiii) sensitize and uncover youth to ethnic and religious prejudices and stereotypes.

Peacebuilding and post-conflict reconstruction in most conflict and post-conflict societies are undertaken by adults or any other agency other that youth. However, although there are attempts to involve young people in the peace building process in recent times, it is not quite widespread as it may be required. In many intra-state conflicts where youth engagement in conflict is high, there is no attempt made to view the problem as centered on youth. Youth are however, assets to peace building process and have the great potential to prevent a relapse in to war or armed conflict.

Engaging youth in the peace building processes have their own share of challenges. Adult leaders often limit the involvement of youth to protect their own positions. These hurdles suggest that understanding youth in the policy and academic discourses and in the context of peace building are difficult. Conflict and peace building are a field of studies that are normative and prescriptive while being descriptive and analytical. Students who are engaged in peace, conflict, development and human rights research are not only interested in why youth engage in violence, they are also motivated with a desire to bring positive change in the lives of young people. They however, have limited opportunity for such critical reflection on the frames and underlying ideologies of their potential. Youth disengagement in the peace building process may arise from naïve ideals, a stubborn and even paternalistic belief in superiority of certain practitioner ideas, and a disregard for young people's ideas. It is often stereotypical of aid workers, teachers and others who have negative views about young people with whom they work. These issues may have structural explanations for this dynamic among people working with young people and can be addressed. These workers are often overburdened, under-resourced and feel marginalized by the higher level decision making processes. These frame, ideological images and debates on youth that have emerged over the last two decades, seem to shape the global policy process. The historical pattern of using children and childhood for military and political ends seem to be directing the path towards the engagement of youth in peace building processes.

In recent years there has been increasing attention on programmes that seek to engage youth in directly in peace building activities and to fully make use of their potential. Such programmes include peace education, training in rights, peace building and conflict resolution and the direct engagement of youth in election and human rights monitoring a d accountability programmes. It is thus important to focus on both active and non-active youth participants in conflict and those who are from war affected regions (Hilker and Fraser, 2009). In the context of north east India, and India's efforts to building sustainable peace in the region the three approaches suggested by Kemper seems appropriate. Kemper's model of youth engagement in peace building suggests (i) a rights based approach, (ii) the economic approach and (iii) the socio-political approach. The rights based approach suggests that it is generally applicable for those who are below the age of 18. The fundamental principle of this approach is the recognition that children are the most affected during conflict and they have certain rights and needs that need to be upheld during such situations (Kemper, 2005). The economic approach primarily focuses on investing on youth in order to help them sustain themselves during conflict situations (Cohen and Arato, 1995). The economic approach asks how the presence of youth affects war and not how war affects youth (Kemper, 2005). Finally the not so often tested approach is the socio-political approach that regards youth as potential contributors to peace and that if their perceptions should be given due importance (Kemper, 2005; Boyden and Mann, 2000). Although all three approaches are different from each other, Kemper suggests that all three approaches are important in order to have a more integrated and holistic approach which would create an enabling environment for youth participation in post conflict situations (ibid). Young people are forced to take part in roles that would generally be considered as those of adults such as war and social upheaval and in order to achieve maximum effectiveness for youth programming, it is imperative to empower youth as decision makers as the process is dominated by adults (Lowicki, 2002). Thus, the need to include youth in the peace building process seems important at this juncture. As we have already noted that conflict in north east India is waning weak, and people in the region have also become conflict weary, I posit that it would be the best time to include youth in the peace building process. The prevailing conditions could with the engagement of youth in peace building processes could result in sustainable peace and the fulfilment of the aspirations and needs of both young people and the adult living in the region.

REFERENCES

Baruah, A.K. (2002) 'Student Power in North East India: Understanding Student Movements', New Delhi: Regency Publications.

Chakrabarti, S.B. (2008) 'Student/Youth Movements with Special Reference to North East India', in Ray, A.K. and Chakrabarti, S.B. (eds.) Society, Politics and Development in North East India: Essays in Memory of Dr. Basudeb Dutta Ray, New Delhi: Concept Publishing Company.

Hilker. L.M. and Fraser, E. (2009) 'Youth Exclusion, Violence, Conflict and Fragile States', Report Prepared for DFID's Equity and Rights Team, London: Social Development Direct.

Koeffel, C. (2003) 'Globalization of Youth Activism and Human Rights', in Fitzroy, A.J (ed) Highly Affected Rarely Considered: The International Youth Parliament Commission's Report on the Globalization of Young People, Australia: Oxfam Community Aid Abroad.

Weiss, M. Aspinall, E. and Thompson, M. (2012) 'Introduction: Understanding Student Activism in Asia in Weiss, M. Aspinall, E. and Thompson, M. (eds.) Student Activism in Asia: Between Protests and Powerlessness, Minneapolis: Minnesota University Press.

Bibliography

Ali, Syed Sajjad 2015. '*Tripura Withdraws AFSPA, Says Insurgency on the Wane*', The Hindu 28 May 2015.

Allen, S. (1968) 'Some Theoretical Problems in the Study of Youth' *The Sociological Review*, 16(3): 319-331.

Apfel, R.J and Simon, B. (1996) '*Introduction*', in Apfel, R.J and Simon, B. (eds.) Minefields in Their Hearts: The Mental Health of Children in War and Communal Violence, New Haven: Yale University Press.

Apfel, R.J. and Simon, B. (1996) '*Minefields in their hearts: The Mental Health of Children in War*', New Haven: Yale University Press.

Ardizzand, C. (2003) 'Generating Peace: A Study of Non-Formal Youth Organizations', *Peace and Change Journal*, 28(3): 420-445.

Argenti, N. (1998) 'Air Youth: Performance, Violence and the State in Cameroon', *Journal of the Royal Anthropological Institute*, 4(4): 753-781.

Argenti, N. (2002) 'Youth in Africa: A Major Resource for Change', in De Waal, A. and Argenti, N. (eds.) *Young Africa: Realizing Rights of Children and Youth*, New Jersey: Africa World Press.

Aries, P. (1965) 'Centuries of Childhood: A Social History of Family Life', New York: Vintage Books.

Asian Development Bank (2012) 'Working Differently in Fragile and Conflict-Affected Situations – the ADB experience: A Staff Handbook', Mandaluyong: Asian Development Bank.

Assad, R. and Deborah, L. (2013) '*Employment for Youth – A growing Challenge for the Global Community*', Background Research Paper of the United Nations High Level Panel on the Post 2015 Development Agenda', Minnesota: University of Minnesota Press.

Baker, G. and Ricardo, L. (2006) 'Young Men and the Construction of Masculinity in Sub-Saharan Africa: Implications for HIV/AIDS, Conflict and Violence', in Bannon, I and Marion, L.L. (eds.) *The Other half of Gender: Men's Issues in Development*, Washington D.C: World Bank.

Banks, M., Bates, I., Brekwell, G., Emler, N., Jameison, L., and Roberts, K. (1992) 'Careers and Identities', Milton Keynes: Open University Press.

Barbora, Sanjay 2008. 'Rethinking India's Counter-Insurgency Campaign in North East', *Economic and Political Weekly*, 41(35): 3805-3812.

Barnett, M., Kam, H., O'Donnell and Sitea (2007) 'Peacebuilding: What's in a Name?', Global Governance, 13(1):35-58.

Baruah, S. (2002) 'Gulliver's Troubles: State and Militants in North East India', Economic and Political Weekly, 37(41): 4178-4182.

Baruah, Sanjib 2005. '*Durable Disorder: Understanding the Politics of North East India*'. New Delhi: Oxford University Press.

Beck, U. (1992) 'Risk Society: Towards a New Modernity', London: Sage Publications.

Behrman, J.R. and Birdsall, N. (1988) 'The Reward for Good Timing: Cohort Effects and Earnings Function for Brazilian Males', *Review of Economics and Statistics*, 70(1): 129-135.

Bennett, A. (2005) 'In Defence of Neo-Tribes: A Response to Blackman and Hesmondhalgh', Journal of Youth Studies, 8(2): 255-259.

Berger, M.C. (1985) 'The Effect of Cohort Size on Earning Growth: A Re-examination of the Evidence', *Journal of Political Economy*', 93(1): 561-573.

Bhaumik, S. (1998) 'North East India: Evolution of Post-Colonial Region', in Chatterjee, P. (ed) Wages of Freedom, New Delhi: Oxford University Press.

Bhaumik, S. (2009) 'Troubled Periphery: Crisis of India's North East', New Delhi: Sage Publications.

Bhaumik, Subir 1998. 'North East India: Evolution of Post-Colonial Region'. In *Wages of Freedom*, edited by Partha Chatterjee'. New Delhi: Oxford University Press. pp. 310-311.

Bhaumik, Subir 2009. '*Troubled Periphery: Crisis of India's North East*', New Delhi: Sage Publications.

Bloom, D.E. and Williamson, J.G. (1998) 'Demographic Transitions and Economic Miracles in Emerging East Asia', *World Bank Economic Review*, 12(3): 419-455.

Bloom, D.E., Freeman, R.B. and Korenman, S. (1987) 'The Labor Market Consequences of Generational Crowding', *European Journal of Population*, 3(1): 131-176.

Boren, M.E. (2001) '*Student Resistance: A History of the Unruly Subject*', New York: Routledge.

Borgohain, H. (1982) 'Manipur: Anatomy of Despair', Economic and Political Weekly, 46(47): 1858-1860.

Boulding, E. (1988) 'Building a Global Civic Culture: Education for an Interdependent World', New York: Teachers College Press.

Bouthal, G. (1970) '*L'infanticide Differe*', Paris: Hatchette.

Boyden, J (1990) 'Childhood and the Policy Makers: A Comparative Perspective on the Globalization of Childhood', in James, Allison and Prout, Alan (eds.) Contemporary Issues in the Sociological Study of Childhood, Philadelphia: Falmer Press.

Boyden, J. and De Berry, J. (2004) 'Introduction', in Boyden, J and De Berry, J. (eds.) Children and Youth on the Frontline: Ethnography, Armed Conflict and Displacement, Oxford: Berghann.

Braungart, R.G. (1984) 'Historical and Generational Patterns of Youth Movements: A Global Perspective', *Comparative Social Research*, 7(1): 3-62.

Brett, R. and McCallin, M. (1998) '*Children: The Invincible Soldiers*', New York: United Nations Children's Fund.

Brown, B.B., Larson, R.W. and Saraswathi, T.S. (2002) '*The World's Youth: Adolescence in Eight Regions of the Globe*', New York: Cambridge University Press.

Bukley-Zistel, S. (2008) 'Conflict Transformation and Social Change in Uganda: Remembering after Violence', Basingstoke: Palgrave MacMillan.

Bureau for Crisis Prevention and Recovery (2005) '*Youth and Violent Conflict: Society and Development in Crisis?*', Geneva: Bureau for Crisis Prevention and Recovery.

Burns, R.J. and Aspeslagh, R. (eds.) (1996) 'Three Decades of Peace Education Around the World: An Analogy', New York: Garland.

Burton, J (1994) 'The Upsurge in Interest in the Relief-Development Continuum: What Does it Mean', Relief and Rehabilitation Network, Newsletter, September 1994.

Butler, J. (2009) '*Frames of War: When is Life Grievable?*', London: Verso.

Chadda, Vivek 2005. '*Low Intensity Conflicts in India: An Analysis*'. New Delhi: Sage Publications.

Chhonkar, Pradip Singh 2016. Naga "Framework Agreement" and its Aftermath', IDSA Comment', New Delhi: IDSA, September 1, 2016. Accessed: 25.10.2016

Choucri, N. (1974) '*Population Dynamics and International Violence: Propositions, Insights and Evidence*', Lexington: Lexington.

Cline, Lawrence E. 2006. 'The Insurgency Environment in North East India', *Small Wars and Insurgencies*, 17(2): 126-147.

Clodfetter, M. (1992) 'Warfare and Armed Conflicts: A Statistical Reference to Casualty and Other Figures 1618-1991', North Carolina: Jefferson McFarland.

Cohen, P. (1997) 'Rethinking the Youth Question: Education, Labour and Cultural Studies', London: Macmillan.

Coles, B. (1995) 'Youth and Social Policy', London: University College London.

Collier, P. and Heoffler, A. (2004) 'Greed and Grievance in Civil War', *Oxford Economic Papers*, 56(1): 563-595.

Conca, K., and Dabelko, G.D. (2018) 'Green Planet Blues: Critical Perspectives on Global Environmental Politics', London: Routledge.

Connell, R.W. (1987) 'Gender and Power', California: Stanford University Press.

Connor W. (1967) 'Self-Determination: The New Phase', World Politics, 20(1): 30-53.

Cordell, K. and Wolff, S (2010) 'Ethnic Conflict', Cambridge: Polity Press.

Das, B. (2010) 'KNLF Rebels Surrender in Assam', http://in.reuters.com/article/idINIndia-46094620100211. Accessed: 17 June 2019.

Das, Biswajyoti (2010) '*KLNLF Rebels Surrender in Assam*', http://in.reuters.com/article/idINIndia-46094620100211. Accessed June 17, 2016.

Das, S.K. (2007) 'Conflict and Peace in North East: The Role of Civil Society', Policy Studies No.42. Washington D.C.: East-West Centre.

Das, Samir Kumar (2007) '*Conflict and Peace in India's North East: The Role of Civil Society*', Policy Studies Brief No. 42, Washington D.C: East West Centre.

Das, Samir Kumar 2010. *India: Democracy Nation and the Spirals of Insecurity: State Response to Ethnic Separatism in India's North East*, in Robert G. Wirsing and Ehsan Ahrari, (ed) 'Fixing Fractured Nations: The Challenge of Ethnic Separatism in the Asia Pacific'. New York: Palgrave Macmillan. pp. 116-139.

Davies, B. (2004) 'Curriculum in Youth Work: An Old Debate in New Clothes?', Youth and Policy, 85(1): 87-98.

Davies, B. (2006) 'the Place of Doubt in Youth Work: A Personal Journey', Youth and Policy 92(1): 69-80.

De Waal, A. (2002) 'Realizing Child Rights in Africa: Children Young People and Leadership', in De Waal, A and Argenti, N (eds.) Young Africa: Realizing Rights of Children and Youth, New Jersey: Africa World Press.

Del Felice, C. and Wisler, A. (2007) 'The Unexplored Power and Potential of Youth as Peace-Builders', Journal of Peace, Conflict and Development, 11(1): 27-45.

Demmers, J. (2012) 'Theories of Violent Conflict: An Introduction', New York: Routledge.

Denskus Tobias (2007) 'Peace Building Does not Build Peace', *Development in Practice*, 17(4/5): 656-662.

Denskus, T. (2007) 'Peacebuilding Does not Build Peace', Development in Practice, 17(5): 656-662.

Dhanesh, R. (2008) 'Youth and Peacebuilding', www.tc.columbia.edu/./Dhaneshyouthandpeacebuilding_22feb08.doc.

Dhillon, G and Yousef, T. (2007) '*Inclusion: Meeting the 100 Million Challenge Flagship Report*. Middle East Youth Initiative: Wolfensohn Center for development at the Brookings Institution and Dubai School of Government.

Dolan, P. (2012) 'Travelling Through Social Support and Youth Civic Action on a Journey Towards resilience', in Ungar, M. (ed) The Social Ecology of Resilience: A Handbook of Theory and Practice, New York: Springer.

Dosse, S. (2010) 'The Rise of Intra-State Wars: New Threats and New Methods', Small Wars Journal, www.smallwarsjournal.com. Accessed: 19.02.2019.

Doyle, Michael W and Sambanis, Nicholas (2000) 'International Peace Building: A Theoretical and Quantitative Analysis', *The American Political Science Review*, 94(4): 779-801.

Drummond-Mundal, L., and Cave, G. (2007) 'Young Peacebuilders: Exploring Youth Engagement with Conflict and Social Change', Journal of Peacebuilding and Development, 3(3): 63-76.

Easterlin, R.A. (1987) 'Easterlin Hypothesis' in Eatwell, J., Millgate and Newman, P. (eds.) *A Dictionary of Economics*, Vol.2. New York: Stockton.

Eberwein, W. and Chojnacki, S. (2001) 'Scientific Necessity and Political Utility: A Comparison of Data on Violent Conflicts', Discussion Paper P01-304, Berlin: Center for Social Science Research.

Edwards, A.R. (1983) 'Sex Roles: A Problem for Sociology and for Women', Journal of Sociology, 19(3): 385-412.

Eisenstadt, S.N. (1956) 'From Generation to Generation: Age Groups and Social Structure', New York: The Free Press.

Elbadawi, I and Sambanis, N. (2000) 'Why are there so Many Civil Wars in Africa?: Understanding and Preventing Conflict', *Journal of African Economies*, 9(3): 244-269.

Elliot, D.S., Hamburg, B.A. and Williams, K.R. (1998) '*Violence in American Schools*', New York: Cambridge University Press.

Emery, M. (2013) 'Social Capital and Youth Development: Toward a Typology of Program and Practices', New Directions for Youth Development, 2013(138): 49-59.

Erickson, E. (1968) 'Identity: Youth and Crisis', New York: Norton Press.

Erickson, E.H. (1968) '*Identity, Youth and Crisis*', New York: Norton.

Esman, M. (1994) 'Ethnic Politics', New York: Cornell University Press.

Etsy, D.C. *et al.* (1998) '*State Failure Task Force Report: Phase II Findings*', Mclean: Science Applications International.

Evans,K. and Furlong, A. (1997) Metaphors of Youth Transitions: Niches, Pathways, Trajectories or Navigations', in Bynner and Furlong, A. (eds.) Youth Citizenship and Social Change in a European Context, Aldershot: Ashgate Publications.

Fares, J., Montenegro, C.E and Orazern, P.F. (2006) '*How are Youth Faring in the Labor Market? Evidence from Around the World*', World Bank Policy Research Paper, 4071.

Farrar, S. and Lowe, J. (1978) 'Sex Role Theory: Political Cul-de-Sac', Refractory Girl, 6(1): 14-16.

Fearon, J.D. and Laitin, D.D. (2003) 'Ethnicity, Insurgency and Civil War', *American Political Science Review*, 97(1): 75-90.

Featherston, Beth (2000) '*From Conflict Resolution to Transformative Peace Building: Reflections from Croatia*', Working Paper 4, Bradford: Centre for Conflict Resolution.

Fetherston, B. (2000) 'From Conflict Resolution to Transformative Peacebuilding: Reflections from Croatia', Working Paper 4. Bradford: Center for Conflict Resolution.

Feure, L.S. (1969) '*The Conflict of Generations: The Character and Significance of Student Movements*', London: Heinemann.

Florquin, N. and Berman, E. (eds.) (2005) '*Armed and Aimless: Armed Groups, Guns and Human Security in the ECOWAS Region*', Geneva: Small Arms Survey.

Freddy, H.J. (2016) 'Engaging Youth: Challenges and Opportunities Towards Building Peace in North East India', Man and Society: A Journal of North East Studies, 13(1): 106-120.

Freddy, Haans J (2016) 'Engaging Youth: Challenges and Opportunities towards Building Peace in North East India', *Man and Society: A Journal of North East Studies*, ICSSR-NERC, 13(1): 106-120.

Freddy, Haans J. 2016. 'Engaging Youth: Challenges and Opportunities towards Building Peace in North East India', *Man and Society: A Journal of North East Studies*, 13(1): 106-120.

Galtung, J. (1969) 'Violence, Peace and Peace Research', Journal of Peace Research, 6(3): 167-191.

Galtung, J. (1996) 'Peace by Peaceful Means: Peace and Conflict Development and Civilization', London: Sage Publications.

Garg, Ibu Sanjeeb 2016. 'Unrest in Tripura-The Struggle and the Way Forward'. The Shillong Times. 6 September 2016.

Ghosh, Atig (2013) '*Governing Conflict and Peace Building in India's North East and Bihar*', Core Policy Brief 08/2013, Oslo: Peace Research Institute Oslo.

Gidden, A. (1991) 'Modernity an Self Identity: Self and Society in the Late Modern Age', Cambridge: Polity Press.

Gillis, J (1994) 'Youth and History: Tradition and Change in European Age Relations 1770-Present', New York Academic Press.

Gladstone J.A. (2001) 'Demography, Environment and Security', in Paul, F.D and Nils, P.G. (eds.) *Environmental Conflict*, Boulder: Westview.

Gladstone, J.A. (1991) '*Revolution and Rebellion in the Early Modern World*', Berkeley: University of California Press.

Goldstein, J. (2001) '*War and Gender: How Gender Shapes the War System and Vice-Versa*', Cambridge: Cambridge University Press.

Goodhand, J and Hulme, D. (1999) 'From Wars to Complex Political Emergencies: Understanding Conflict and Peacebuilding in the New World Disorder', Third World Quarterly, 28(1): 13-26.

Goodhand, Jonathan and Hulme, David (1999) 'From Wars to Complex Political Emergencies: Understanding Conflict and Peace Building in the New World Disorder', *Third World Quarterly*, 20(1): 13-26.

Goswami, Namrata (2011) '*A Non-Territorial Resolution to the Naga Conflict*' IDSA Strategic Comment, Nov, 15, 2011. Accessed, June 15, 2016.

Goswami, Namrata (2015) '*The Naga Peace Accord: Why Now?*', IDSA Strategic Comment, August 7, 2015. Accessed June 15, 2016.

Goswami, Namrata 2009. '*An Assessment of Insurgencies in Assam, Manipur and Nagaland in 2009*'. IDSA Issue Brief. New Delhi: Institute of Defense and Strategic Analysis.

Goswami, Namrata 2011. '*A Non-Territorial Resolution to the Naga Ethnic Conflict*', IDSA Comment, New Delhi: Institute of Defense and Strategic Analysis.

Goswami, U. (2014) 'Conflict and Reconciliation: The Politics of Ethnicity in Assam', New Delhi: Routledge.

Gurr, T. (1970) 'Why Men Rebel', Princeton: Princeton University Press.

Hajer, M.A and Wagenaar (2003) 'Deliberative Policy Analysis: Understanding Governance in the Network Society', Cambridge: Cambridge University Press.

Hall, G.S. (1904) 'Adolescence: Its Psychology and its Relation to Physiology, Anthropology, Sociology, Sex, Crime, Religion and Education', New York: Appleton and Co.

Hall, T.D. (2004) 'Ethnic Conflicts as a Global Social Problem in George Ritzer (ed) 'Handbook of Social Problems: A Comparative International Perspective', Thousand Oaks; Sage Publications.

Haokip, Thangkholal 2013. The Kuki-Naga Conflict in the Light of Recent Publications', *South Asia Research*, 33(1): 77-87.

Harbom, L. and Wallensteen, P. (2009) 'Armed Conflicts 1946-2008', Journal of Peace Research, 46(4): 577-587.

Harff, B. and Gurr, T. (1988) 'Toward Empirical Theory of Genocide and Politicides: Identification and Measurement of Cases Since 1945', International Studies Quarterly, 32(3): 359-371.

Harris, I. (2012) 'Peace Education theory', Journal of Peace Education, 1(1): 5-20.

Harris, I. and Morrison, M.L. (2003) 'Peace Education', Jefferson: McFarland.

Hassan, Sajjad 2008. 'Understanding the Breakdown of North East: Identity Wars or Crises of Legitimacy', *Journal of South Asian Development*, 3(1): 53-86.

Hassan, Sajjad M. 2007. '*Understanding the Breakdown of North East India: Explanations in State-Society Relations*', London School of Economics, Working Paper SeriesNo. 07-83. London: London School of Economics.

Head, B.W. (2011) 'Why not Ask Them? Mapping and Promoting Youth Particiaption', Children and Youth Services Review, 33(4): 541-547.

Heinsohn, G. (2003) '*Sons and World Power: Terror in the Rise and Fall of Nations*', Bern: Orell Fusli.

Hendrixson, A. (2004) '*Angry Young Men, Veiled Young Women: Constructing a New Population Threat*', Corner House Briefing No. 34.

Hindustan Times (2010) '*ULFA Chairman Rajkhowa Held in Bangladesh, Flown to Delhi*', Hindustan Times, December 9, 2009, http://www.hindustantimes.com/ULFA-Chairman-Rajkhowa-held-in-Bangladesh-flown-to-Delhi/Article1-482481.aspx. Accessed June 17, 2016.

Hobsbawm, E. (1994) 'Age of Extremes: The Short Twentieth Century 1914-1991', London: Abacus.

Holmes, J.S. (2001) '*Radicalism: Is the Devil in the Demographics?*', New York Times, December 9, 2001.

Horowitz, D.L. (1985) 'Discourses on Violence: Conflict Analysis Reconsidered', Manchester: Manchester University Press.

Hudson, V.M. and Den Boer, A.M. (2004) '*Bare Branches: The Security Implications of Asia's Surplus Male Population*', Cambridge: MIT Press.

Hughes, P.M. (1997) '*Global Threats and Challenges to the United States and its Interests Abroad: Statement for the Senate Select Committee on Intelligence*', February 6, 1997, Washington D.C: Defence Intelligence Agency.

Huntington, S.P. (1996) '*The Clash of Civilizations and the Remaking of the World Order*', New York: Simon and Schuster.

Hussain, W. (2008) 'The ULFA Mutiny', Outlook India', July 03, 2008, http://outlookindia.com/article.aspx/237821. Accessed: 17 June 2019.

Hussain, Wasbir (2008) '*The ULFA Mutiny*', Outlook India July 03, 2008, http://www.outlookindia.com/article.aspx? 237821. Accessed June 15, 2016.

Innoue, K. (2005) 'Integration of North East: The State Formation Process, in Sub-Regional Relations in Eastern South Asia: With Special Focus on India's North Eastern Regions' Study No. 133. Tokyo: Institute for Developing Economies.

Inoue, Kyoko 2005. '*Integration of North East: The State Formation Process. In Sub-Regional Relations in Eastern South Asia: With Special Focus on India's North Eastern Region*'. Study No.133. Tokyo: Institute of Developing Economies.

International Labor Organization (2013) '*Global Employment Trends for Youth 2013: A Generation at Risk*', Geneva: International Labor Organization.

Irwin, S. (1995) 'Social Change and the Transition from Youth to Adulthood', London: UCL Press,

Jackson, N. and Flemingham (2003) '*The Demographic Gift in Australia*', Discussion Paper, School of Economics, Hobart: University of Tasmania.

Jenkins, Rob (2013) '*Peace Building: From Concept to Commission*', New York: Routledge.

Jenskins, R. (2013) 'Peacebuilding and State Building Priorities and Challenges: A Synthesis of Findings from Seven Multi-Stakeholder Consultations', Dili, Timor-Leste: Organization for Economic Cooperation and Development.

Johal, R.K., Kaur, J., Begra, S., and Manchanda, P.K. (2012) 'Situation Analysis on Youth and Local Governance', Chandigarh: Commonwealth Youth Programme, Asia Centre.

Johnson, D.W and Johnson, R.T. (2006) 'Peace Education for Consensual Peace: The Essential Role of Conflict Resolution', Journal of Peace Education, 3(2):147-174.

Jones, G, and Wallace, C. (1992) 'Youth, Family and Citizenship', Buckingham: Open University Press.

Jones, G. (1988) 'Integrating Process and Structure in the Concept of Youth: A Case for Secondary Analysis', The Sociological Review, 36(4):706-732.

Jurgensmeyer, M. (1993) 'The New Cold War? Religious Nationalism Confronts the Secular State', Berkeley: University of California Press.

Kaplan, R.D. (1994) 'The Coming Anarchy', *Atlantic Monthly*, 273(1): 44-76.

Kaplan, R.D. (1996) '*The Ends of the Earth: A Journey at the Dawn of the 21st Century*', New York: Random House.

Kemper, Y. (2005) 'Youth in War-to-Peace Transitions', Berghof Center for Constructive Conflict Management Report No. 10. The Netherlands: Berghof Center for Constructive Conflict Management.

Kennedy, P. (2003) '*Europe's Laggards Will Never Balance U.S. Power*', The Guardian, June 24, 2003.

Kipgen, Nehginpao 2013. '*Intricacies of Kuki and Naga Ethnocentricism in Manipur*', www.huffingtonpost.com/nehginpao-kipgen/intricacies-of-kuki-and-naga_b_2531115.html. Accessed: 25.10.2016.

Kipgen, Nehginpao 2015. '*Manipur on the Brink*', The Hindu, September 2, 2015. Accessed: 25.10.2016.

Klingelhofer, Stephan and Robinson, David (2001) '*The Rule of Law, Custom and Civil Society in the South Pacific in the 21st Century: Challenges and Opportunities*', Washington D.C: International Center for Not-for-Profit Law.

Kolas, Ashlid 2015. 'Framing the Tribal: Ethnic Violence in North East India', *Asian Ethnicity*, http://dx.doi.org/10.1080/14361369.2015.1062050. Accessed: 17.10.2016.

Korenman, S. and Neuman, D. (2000) 'Cohort Crowding and Youth Labor Markets: A Cross-Sectional Analysis', in Blanchflower, D. and Freeman, R. (eds.) *Youth Employment and Joblessness in Advanced Countries*, Chicago: Chicago University Press.

Kreiberg, Louis 2011. 'The State of the Art in Conflict Transformation', in Beatrix Austin, Martin Fischer, Hans J. Giessman (ed) *Advanced Conflict Transformation: The Berghof Handbook II*. Berlin: Berghof Foundation. pp. 49-69.

La Graffe, D. (20102) 'The Youth Bulge in Egypt: An Intersection of Demographics, Security and the Arab Spring', *Journal of Strategic Security*, 5(2): 65-79.

Lacina, B. (2009) 'The Problem of Political Stability in North East India: Local Ethnic Autocracy and the Rule of Law', Asian Survey, 49(6): 998-1020.

Ladwig, W. (2009) 'Managing Separatist Insurgencies: Insights from North Eastern India', Paper Presented at the International Studies Association Conference, New York, 15-18 February 2009.

Ladwig, Walter III (2009) '*Managing Separatist Insurgencies: Insights from North Eastern India*', Paper Prepared for the International Studies Association Conference, New York, 15-18 February 2009.

Ladwig, Walter III (2009) '*Managing Separatist Insurgencies: Insights from North Eastern India*', Paper Prepared for the International Studies Association Conference, New York, 15-18 February 2009.

Lake, D.A. and Rothchild, D. (1998) (eds.) 'The International Spread of Ethnic Conflict', Princeton: Princeton University Press.

Lam, D. (2011) 'How the World Survived the Population Bomb: Lessons from Fifty Years of Extraordinary Demographic History, *Demography*, 48(4): 1231-1262.

Lam, D. and Leticia, M. (2008) 'Stages of the Demographic Transition from and Child's Perspective: Family Size, Cohort Size and Schooling', *Population and Development Review*, 34(2): 225-252.

Lauder, H., Brown,P., Dillabough, J. and Haley, A.H. (eds.) (2007) 'Education, Globalisation and Social Change', Oxford: Oxford University Press.

Lee, R. (2003) 'The Demographic Transition: Three Centuries of Fundamental Change', *Journal of Economic Perspectives*, 17(1): 167-190.

Liebau, E. and Chisholm (1993) 'Youth, Social Change and Education: Issues and Problems', Journal of Education Policy, 8(1):3-8.

Llamazares, Monica (2005) '*Post War Peace Building Reviewed: A Critical Exploration of Generic Approaches to Post-War Reconstruction*', Working Paper 14, Bradford: Centre for Conflict Resolution.

Lowickci, J. and Pillsbury, A. (2000) '*Untapped Potential: Adolescents Affected by Armed Conflict: A Review of Programs and Policies*', New York: Women's Commission for Refugee Women and Children.

Machel, G. (2001) '*The Impact of War on Children: A Review of the Progress Since 1996*', Geneva: United Nations.

Mahanta, N.G. (2013) 'Confronting the State: The ULFA's Quest for Sovereignty', New Delhi: Sage Publications.

Mannheim, K. (1952) 'The Problem of Generations', in Mannheim, K. Essays on the Sociology of Knowledge, London: Routledge and Kegan Paul.

Marshall, T.H. and Bottomore, T. (1992) 'Citizenship and Social Class', London: Pluto Press.

Mathur, A. (2011) 'A Winning Strategy for India's North East', Jindal Journal of International Affairs, 1(1): 269-298.

Mathur, Akshay (2011) 'A Winning Strategy for India's North East', *Jindal Journal of International Affairs*, 1(1): 269-298.

Maytok, T., Senehi, J. and Byrne, S. (2011) '*Critical Issues in Peace and Conflict Studies: Theory, Practice and Pedagogy*', Maryland: Lexington Books.

Mazumdar, A. (2005) 'Bhutan's Military Action Against Indian Insurgents', Asian Survey, 45(4): 566-580.

Mazumdar, Arijit (2005) 'Bhutan's Military Action against Indian Insurgents', *Asian Survey*, 45(4): 566-580.

McEvoy-Levy, S. (2001) 'Youth as Social and Political Agents: Issues in Post-Settlement Peacebuilding', Kroc Institute Occasional Paper #21:OP2.

Midlarski, M. (1988) 'Rulers and Ruled: Patterned Inequality and the Onset of Mass Political Violence', American Political Science Review, 82(2): 491-509.

Miller, M. (2013) 'Studies in Asian Security: Wronged by Empire: Post-Imperial Ideology and Foreign Policy in India and China', California: Stanford University Press.

Miller, Manjari 2013. '*Studies in Asian Security: Wronged by Empire: Post-Imperial Ideology and Foreign Policy in India and China*,' California: Stanford University Press.

Misra, U. (2000) 'The Periphery Strikes Back: Challenges to the Nation-State in Assam and Nagaland', Shimla: Indian Institute of Advanced Study.

Moller, B. (2003) 'Conflict Theory', Working Paper Series, 122, Denmark – Aalbory: DIR and Institute for History, International and Social Studies.

Moller, S.H. (1968) 'Youth as a Force in the Modern World', *Comparative Studies in Society and History*', 10(1): 238-260.

Mukherjee, J.R. (2005) 'Insiders Experience of Insurgency in India's North East', London: Anthem Press.

Mungham, G. and Parsons, G. (eds.) (1978) 'Working Class Youth Culture', London: Routledge and Kegan Paul.

Neopolitan, J.L. (1997) '*Cross-National Crime: A Research Review and Sourcebook*', Westport: Greenwood.

Neumayer, E. (2003) 'Good Policy Can Lower Violent Crime: Evidence from a Cross-National Panel of Homicide Rates 1980-97', *Journal of Peace Research* 40(1): 247-262.

Niang, S.R (2010) '*Terrorizing Ages: The Effects of Youth Densities and the Relative Youth Cohort Size on the Likelihood and the Pervasiveness of Terrorism*', Paper Presented at the Annual Meeting of the Midwest Political Association, April 2010, Chicago.

OECD (2011) '*Rethinking the Involvement of Youth in Armed Violence: Programming Note*', http://dx.doi.org/10.1787.9789264107205-en.

Ozerdem, Alpaslan and Lee, Sung Yong (2015) '*International Peace Building: An Introduction*', London: Routledge.

Parsons, T. (1942) 'Age and Sex in the Social Structure of the United States', American Sociological Review, 7(1): 604-616.

Phanjoubam, P. (2016) 'The North East Question: Conflict and Frontiers', New York: Routledge.

Phanjoubam, Pradip 2016. '*The North East Question: Conflicts and Frontiers*'. New York: Routledge.

Phukan, G. (1996) 'Politics of Regionalism in North East India', New Delhi: Spectrum Publications.

Piang, Lam Khan L. 2015. 'Overlapping Territorial Claims and Ethnic Conflict in Manipur', *South Asia Research* 35(2): 158-176.

Prabhakara, M.S. (2007) 'Separatist Movements in North East India: Rhetoric and Reality', Economic and Political Weekly, 42(9): 728-730.

Pysnakova, M. and Miles, S. (2010) 'The Post-Revolutionary Consumer Generation: Mainstream Youth and the Paradox of Choice in the Czech Republic', Journal of Youth Studies, 13(5): 533-547.

Rajagopalan, Swarna (2008) '*Peace Accords in North East India: Journey Over Milestones*', Washington D.C.: East West Centre.

Rajiv Gandhi National Institute of Youth Development (2012) 'Youth Development Report 2010' Sriperumbudur: Rajiv Gandhi National Institute of Youth Development.

Reardon, B. (1993) 'Women and Peace: Feminist Visions of Security', New York: SUNY Press.

Reardon, B. (1998) 'Comparative Peace Education: Educating for Global Responsibility', New York: Teachers College Press.

Reardon, B. (2001) Education for a Culture of Peace in a Gender Perspective, Paris: UNESCO.

Reynolds, P. (1998) 'Activism, Politics and the Punishment of Children', in Van Buren, G. (ed) Childhood Abused, Hampshire: Ashgate.

Ropers, Norbert 2008. '*Systemic Conflict Transformation: Reflections on the Conflict and Peace Process in Sri Lanka*'. Berlin: Berghoff Foundation.

Routray, Bibhu Prasad 2011. '*Conflict Transformation in Assam: Lesson and Challenges*', www.claws.in/621/conflict-transformation-in-assam-lessons-and-challenges-bibhu-prasad-routray.html. Accessed: 25.10.2016.

Rowe, R.C et.al., (2004) 'Testosterone, Anti-Social Behavior and Social Dominance in Boys: Pubertal Development and Bio-Social Interaction', *Biological Psychiatry*, 55(1):546-552.

Sahaya D.N. 2011. 'How Tripura Overcame Insurgency', The Hindu. 19 September 2011.

Sandhan, O. (2005) 'Who Rules Manipur's Streets?', Article No, 1839. New Delhi: Institute of Peace and Conflict Studies.

Schomaker, R. (2013) 'Youth Bulges, Poor Institutional Quality and Missing Migration Opportunity – Triggers of the Potential Counter-Measures for Terrorism in MENA', *Topics in Middle Eastern and African Economies*, 15(1): 116-140.

Schwartz, S. (2008) 'The Dynamic Role of Youth in Post-Conflict Reconstruction: Lessons from Mozambique, The Democratic Republic of Congo and Kosovo', Unpublished Theses, Wesleyan University. www.un.org/./peacebuilding/.Guiding per cent 20principles per cent 20Youth per cent 20pa.cached.

Sethi, P.C. (1983) 'Home Minister of India Statement in Lok Sabha, 14 March 1983', in Documents of Assam, Part B, edited by Trivedi, V.R. New Delhi: Omsons Publications.

Shimray, U.A. (2004) 'Socio-Political Unrest in the Region Called North East India', Economic and Political Weekly, 39(42): 4737-4643.

Singh, B.P. (1987) 'North East India: Demography, Culture and Identity Crisis', Modern Asian Studies, 21(2): 257-282.

Siobhan, M. (2011) 'Children, Youth and Peacebuilding', in Matyok, T., Senehi, J., Byrne, S (eds.) Critical Issues in Peace and Conflict Studies: Theory, Practice and Pedagogy, London: Lexington Books.

Sirinate, Vasundhra 2015. 'Can an Accord End an Insurgency?', *The Arena*, The Hindu Centre for Politics and Public Policy, August 7, 2015. www.thehinducentre.com/the-arena/current-issue/article7511878. Accessed: 25.10.2016.

Small, M. and Singer, J.D. (1982) 'Resort to Arms: International and Civil War, 1916-1980', California: Sage Publications.

Social Development Direct (2009) '*Youth Exclusion, Violence, Conflict and Fragile States*', London: Social Development Direct.

Sommerfelt, O.H. and Vambheim, V. (2008) 'The Dream of the Good – A Peace Education Project Exploring the Potential to Educate for Peace at the Individual Level', Journal of Peace Education, 5(1): 79-96.

Sommers, M (2001) '*Youth Care and Protection of Children in Emergencies: A Field Guide*', New York: Save the Children.

Sommers, M. (2006) 'Youth and Conflict: A Brief Review of the Available Literature', Washington, D.C: USAID.

Srikanth, H. and Thomas, C.J. (2005) 'Naga Resistance Movement and the Peace Process in North East India', Peace and Democracy in South Asia, 1(2): 57-87.

The Hindu 2016. '*NSCN has not Given up on Sovereignty, Says Muivah*', The Hindu, New Delhi, 8 July 2016. Accessed: 25.10.2016.

Thorup, C.I. and Kinkade, S. (2005) 'What Works in Youth Engagement in the Balkans', Baltimore: International Youth Foundation.

Times of India (2010) '*Bangladesh Hands Over NDFB Chief, Ranjan Daimary to BSF*' Times of India May 01, 2010, http://timesofindia.indiatimes.com/india/Bangladesh-hands-over-NDFB-chief-Ranjan-Daimary-to-BSF/articleshow/5879908.cms. Accessed June 19, 2016.

Times of India (2011) '*India, Myanmar to Sign Legal Aid Agreement*', http://timesofindia.indiatimes.com/India-Myanmar-to-sign-legal-aid-agreement/articleshow/7158194.cms. Accessed June 17, 2016.

Times of India 2010. '*Bangladesh Hands Over NDFB Chief Ranjan Daimary to BSF*'. Times of India, May 1, 2010.

Toh, S.H. and Cawagas, V.F. (1991) 'Peaceful Theory and Practice in Value Education', Quezon City: Pheonix.

Tschirgi, N. (2004) 'Post-Conflict Peacebuilding Revisited: Achievements, Limitations, Challenges', New York: International Peacebuilding Academy.

Tungpo, Siamchinthang 2016. '*Manipur is Heading for Ethnic Conflict*', Imphal Times June 5, 2016. Accessed: 25.10.2016.

UNDP (1994) 'Human Development Report 1994', Oxford: Oxford University Press.

UNICEF (2007) '*Children and Conflict in a Changing World*', New York: UNICEF.

United Nations (2004) 'Humanitarian Appeal 2004', www.un.org/depts/ocha/cap/appeals.html. Accessed: 15.02.2019.

United Nations (2007) 'United Nations General Assembly Resolution 50/81 of 13 March 1996', The World Programme of Action for Youth in the Year 2000 and Beyond, A/Res/50/81.

United Nations (2010) 'United Nations Peace Building: An Orientation', New York: United Nations Peace Building Support Office.

United Nations Department of Economic and Social Affairs (2005) 'World Youth Report 2005: Young People Today and in 2015', New York: United Nations.

United Nations Environment Programme (2009) 'From Conflict to Peacebuilding: The Role of Natural Resources and the Environment', Nairobi: United Nations Environment Programme.

United Nations General Assembly (2004) 'United Nations Secretary-General's High-Level Panel on Threats, Challenges and Change', New York: United Nations General Assembly.

United Nations Peacebuilding Support Office (2012) 'Informal Brainstorming Consultation on Youth Participation in Peacebuilding Report', New York: United Nations.

United Nations Secretary General (2012) 'Peacebuilding in the Aftermath of Conflict', 1/67/499.

United Nations Security Council (2007) 'Statement 2007/2 by the President of the Security Council', New York: United Nations Security Council.

Upadhyay, Archana 2006. 'Terrorism in the North East: Linkages and Implications', *Economic and Political Weekly*, 41(48): 4993-4999.

Updhyay, A. (2006) 'Terrorism in North East: Linkages and Implications', Economic and Political Weekly, 41(48): 4993-4999.

Urdal, H. (2004) '*The Devil in the Demographics: The Effect of Youth Bulges on Domestic Armed Conflict, 1950-2000*', Social Development Papers: Conflict Prevention and Reconstruction Paper No. 14. Washington D.C: World Bank.

Urdal, H. (2006) 'A Clash of Generations? Youth Bulge and Political Violence', *International Studies Quarterly*, 50(1): 607-629.

USAID (2006) '*Youth and Conflict: A Brief Review of the Available Literature*', New York: USAID.

Vadlamannati, K.C. (2011) "Why Indian Men Rebel? Explaining Armed Rebellion in the North Eastern States of India 1970-2007', Journal of Peace Research, 48(5): 605-619.

Vadlamannti, Krishna Chaitanya 2011. 'Why Indian Men Rebel? Explaining Armed Rebellion in the North Eastern States of India 1970-2007', *Journal of Peace Research*, 48(5): 605-619.

Vidisha, B. (2006) 'Terrorism in India', Huntsville: Sam Houston State Univesity.

Wallace, C and Cross, M. (eds.) (1990) 'Youth in Transition: The Sociology of Youth and Youth Policy', London: Falmer Press.

WELCH, F. (1979) 'Effects of Cohort Size on Earnings: The Baby Boom Babies, Financial Bust', *Journal of Political Economy*, 8(5): 65-97.

World Bank (2007) '*World Bank Report*', Washington D.C: The World Bank.

World Bank (2011) '*Children and Youth in Conflict*', New York: World Bank.

World Health Organization (2002) '*World Report on Health and Violence*', Geneva: World Health Organization.

World Health Organization (2004) '*Preventing Violence and Reducing its Impact: How Development Agencies Can Help* ', Geneva: World Health Organization.

Yilmaz, M.E. (2007) 'Intra-State Conflicts in the Post-Cold War Era', International Journal on World Peace, 24(4): 11-33.

Zakaria, F. (2001) '*The Roots of Rage*', Newsweek, 138(1): 14-33.

Zimmerman, K.F. (1991) 'Ageing and the Labor Market: Age structure, Cohort Size and Unemployment', *Journal of Population Economics*, 4(1): 177-200.

Index

www.ingramcontent.com/pod-product-compliance
Ingram Content Group UK Ltd.
Pitfield, Milton Keynes, MK11 3LW, UK
UKHW021954270726
14060UKWH00002B/511